Turbulence

Frans T.M. Nieuwstadt · Bendiks J. Boersma
Jerry Westerweel

Turbulence

Introduction to Theory and Applications of Turbulent Flows

 Springer

Frans T.M. Nieuwstadt
J.M. Burgers Center for Fluid Mechanics
Delft University of Technology
Delft
The Netherlands

Jerry Westerweel
J.M. Burgers Center for Fluid Mechanics
Delft University of Technology
Delft
The Netherlands

Bendiks J. Boersma
J.M. Burgers Center for Fluid Mechanics
Delft University of Technology
Delft
The Netherlands

Original Dutch text translated by Just van der Wolf.

Frans T.M. Nieuwstadt is deceased.

ISBN 978-3-319-81075-1   ISBN 978-3-319-31599-7   (eBook)
DOI 10.1007/978-3-319-31599-7

Printed on acid-free paper

This Springer imprint is published by Springer Nature
The registered company is Springer International Publishing AG Switzerland

The original version of this book was revised:
An Erratum to this can be found
at DOI 10.1007/978-3-319-31599-7_11

In memoriam,
Frans T.M. Nieuwstadt
April 8, 1946–May 18, 2005

Photo by Sam Rentmeester / FMAX

Preface

Turbulence is a part of fluid mechanics. Therefore, in this book, it is assumed that the reader is already familiar with the fundamentals of fluid mechanics. There are many books that the uninformed reader can consult. A first introduction is provided by the book of White (2011), which also can be recommended for its practical approach, or the books by Acheson (1990), Faber (1995), or Kundu and Cohen (2004), which can be recommended to the reader interested in the physical aspects of fluid mechanics. Standard text books on fluid mechanics are those by Landau and Lifshitz (1959), and especially by Batchelor (1967), which contains a solid mathematical treatise of fluid mechanics.

We are only able to dwell briefly on the results of linear stability theory and the solution to Burgers equation. For further details on stability analysis the reader is referred to the book of Drazin and Reid (1981), while for the Burgers equation the reader is referred to the book of Whitham (1974), where a comprehensive treatise on the Burgers equation is given. An introduction to nonlinear dynamical systems and chaos theory is given in the books by Schuster (1984) and Bergé et al. (1984).

On the topic of turbulence, several text books can be recommended to be used simultaneously with this book; it is often very clarifying when the same material is considered from different viewpoints. Foremost, we suggest the book of Tennekes and Lumley (1972), which has been among the most cited books on turbulence for decades and which has been the inspiration for certain parts of the present book. Traditional descriptions of turbulence that originate from statistical mechanics can be found in the books by Monin and Yaglom (1973) and Landahl and Mollo-Christensen (1986). Also, there are a number of standard works in the field of turbulence, which can be consulted for various topics. Classic text books on turbulence are those by Townsend (1976) and Hinze (1975), while more recent books are those by Pope (2000) and Davidson (2004), which all can be used by those who wish to continue on the topics introduced in this book. There are also several books on specialized topics in turbulence, such as the book by Batchelor (1953) on the theory of homogeneous turbulence, while developments in the field of spectral models can be found in the book by Lesieur (2008) and in the field of renormalization methods in the book by McComb (1990). Also, much attention in physics has been devoted recently to scaling of the microstructure, following the theory of Kolmogorov. An overview of this modern theory can be found in the book by Frisch (1995).

This book was originally written by Frans T.M. Nieuwstadt to support his lectures on turbulence at the level of master students at the Delft University of Technology. It was based on his lecture notes for a course taught at the University of Utrecht before he was appointed at the Delft University of Technology. His objective had been to write a concise introduction on the physical aspects of turbulence (partly inspired by the work of Tennekes and Lumley), but substantially extended to include insights from nonlinear dynamical systems and chaos theory, stability analysis, modern numerical methods, and an overview of current turbulence *closure models* used in *computational fluid dynamics* (CFD) codes. Besides, he wanted to have a book that was also affordable to students.

The original work was written in Dutch, and it was used also at other Dutch universities. However, since around the year 2000, courses had to be taught in English, and we resorted to English language textbooks. Although various excellent books have been available, we could not find the mix of topics that we were used to in the original book by Frans T.M. Nieuwstadt. Since long we had planned to translate, update, and extend the book. Also, we received requests from colleagues to make available a translation of the book. The present book is the result of this effort.

We are indebted to many colleagues who contributed to the completion of this book; in particular we would like to thank Gijs Ooms for proofreading this book and Herman Clercx of the *Vortex Dynamics and Turbulence Group* at the Eindhoven University of Technology for writing a special topic on *rotating turbulence* (Sect. B.4).

Delft   Bendiks J. Boersma
April 2015   Jerry Westerweel

Contents

About the Authors

Frans T.M. Nieuwstadt (1946–2005) was director of the *Laboratory for Aero and Hydrodynamics at the Delft University of Technology* from 1986 till 2005. Previously he worked at the *Royal Netherlands Meteorological Institute* (KNMI), where he conducted research on the atmospheric boundary layer. After completing his studies in Aeronautics in Delft, he worked for 2 years under the supervision of Anatol Roshko at the *California Institute of Technology*. At KNMI he obtained his Ph.D. under the supervision of Henk Tennekes, Jakob Steketee and Jeff Zimmerman. In Delft his interest in turbulence expanded to various areas, such as polymer drag reduction, transition to turbulence in pipe flow, disperse multiphase turbulent flows, and turbulent reacting flows, while maintaining an interest in atmospheric turbulence. He was one of the initiators of the *J.M. Burgers Centre*, which is the Netherlands research school for fluid mechanics that encompasses the activities of entire fluid mechanics community in the Netherlands. Also he was chairman of the *Foundation for Fundamental Research on Matter* (FOM) from 2000 till 2005. He was one of the founding editors of the scientific journal *Flow, Turbulence & Combustion*.

Bendiks J. Boersma (1969–) studied mechanical engineering at the *University of Twente* under supervision of Leen van Wijngaarden. He obtained his Ph.D. at the *Delft University of Technology* in 1997 under supervision of Frans Nieuwstadt. His main interests are the numerical simulation of turbulent flows, including aeroacoustics, drag reduction, and supercritical fluids. After his Ph.D. he worked for 2 years at the *Center for Turbulence Research at Stanford University*, and then became a *Research Fellow* with the *Royal Netherlands Academy of Arts and Sciences*. He was appointed full professor in 2007 at the *Delft University of Technology*, and currently leads the *Energy Technology* section.

Jerry Westerweel (1964–) studied applied physics at the *Delft University of Technology*. He obtained his Ph.D. in 1993 under supervision of Frans Nieuwstadt. As a *Research Fellow* with the *Royal Netherlands Academy of Arts and Sciences* he *worked at Stanford University*, the *California Institute of Technology*, and the *University of Illinois at Urbana-Champaign*. He became an *Anthony van Leeuwenhoek* professor at the *Delft University of Technology* in 2002, and leads the *Fluid Mechanics* section since 2005. His scientific interests are turbulence and coherent flow structures, and optical measurement techniques for quantitative measurements in flows.

Chapter 1
Introduction

Turbulent flows are omnipresent in nature and in technology. In technology, turbulent flows occur for example in nozzles and pipes, followed closely by the flow in devices such as heat exchangers, combustion engines, and turbo machinery. Also, turbulent flow is almost always observed around moving objects, such as airplanes, trains and cars, influencing the flow resistance of those bodies. At the same time, turbulence plays an important part in a variety of transport phenomena, such as heat and mass transfer, but also in flow-induced mixing. An extra complication in this last case is the occurrence of chemical reactions during the mixing, as is the case in combustion. In nature, turbulence plays a part in flows on a geophysical scale, such as the flows in the atmosphere and in the ocean. Transport phenomena in our atmosphere are, for example, mainly controlled by turbulence. An example of this is the distribution of air pollution by turbulent diffusion. Also, people traveling by plane may experience turbulence at first hand by vigorous agitation of the aircraft. On a somewhat larger scale, our weather and even our climate could be called a turbulent phenomenon. Lastly, turbulence is not restricted to our planet, but also plays an important part in flows occurring in the photosphere of stars, the closest being our own sun, and in the formation of planets in accretion discs.

The interest in turbulent flows has increased considerably in the last several decades. There are two main reasons for this. First, turbulence remains an unsolved problem from both a physical and a mathematical point of view. Second, in many practical flow problems, it appears that an inadequate model of turbulence is the most obstructive factor to a solution of the problem at hand.

In a first introduction to fluid mechanics, the concept of *laminar flow* is often introduced immediately, which is then followed by the definition of a turbulent flow as 'non-laminar' flow. Let us compare the characteristics of both types of flow:

laminar	turbulent
layered, regular	disordered
smooth	fluctuating
ordered	chaotic

© Springer International Publishing Switzerland 2016
F.T.M. Nieuwstadt et al., *Turbulence*, DOI 10.1007/978-3-319-31599-7_1

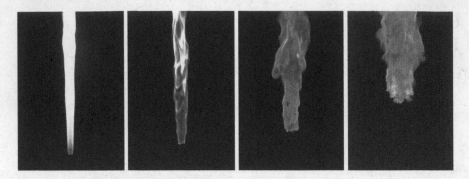

Fig. 1.1 Laminar and turbulent flames. In the laminar flame the combustion is limited by diffusion, whereas in the turbulent flames the combustion is much stronger due to mixing, which results in a higher flame temperature. The laminar flame ($Re \sim 400$) is smooth and has no evident flow structures, whereas the turbulent flames (increasing to $Re \sim 4000$) display a disordered structure and chaotic motion which continuously changes. Images courtesy of: Luis Arteaga and Mark Tummers

On the basis of common experience, almost everyone is somewhat familiar with these qualitative characteristics. An example is shown in Fig. 1.1 for laminar and turbulent flames. The differences between the two types of flow might however lead to the idea that turbulent flow obeys different equations of motion than laminar flow. This idea is not confirmed by experiments, and nowadays there is no doubt that both types of flow obey the same equations of motion. On this basis, we can now ask a simple question: What is the essence of turbulence, and how can we understand turbulent flow as a solution to the equations of motion?

The dynamics of liquids and gases can be described by the laws published by Isaac Newton in 1687 in his *Principia*. At first, Newton devised these equations for the mechanics of solid bodies. However, in subsequent years these laws were extended to frictionless fluids by, among others, Euler and Bernoulli. The formulation of the complete set of equations of motion for a fluid, including flow with friction, did not emerge until the middle of the 19th century. These are the *Navier–Stokes equations* that have the form of a system of nonlinear partial differential equations, describing the relation between the variables of flow, such as velocity and pressure, as a function of position and time.

The Navier–Stokes equations are not sufficient to completely determine the flow in a defined volume. For this we need to specify additional conditions. These conditions determine what the flow on the volume boundaries should look like, that is the so-called *boundary conditions*. Additionally, we need to know the so-called *initial condition*, which is the complete flow as a function of position at an initial moment.

The essential aspect of Newton's laws, and thus of the Navier–Stokes equations, is that they are *deterministic*. This means that *in principle*, given the equations of motion together with the initial and boundary conditions, the evolution of the flow field can be computed as a function of time; hence, the solution to the equations and conditions that describe the flow is completely determined. In other words, the deterministic character of Newton's mechanics implies full *predictability* of the fluid motion.

This can be regarded as a philosophical world view; the mathematician Pierre-Simon Laplace elaborated on this in detail, which was laid down in his work *Mécanique Céleste* that appeared in three volumes between 1799 and 1825. Allegedly, Laplace stated:

Give me the velocity and position of every molecule, and I will predict your future.

In the 19th century the mathematician Hadamard formulated the notion of a *well-posed problem*. This means that a problem is well-posed when the solution of a set of differential equations obeys the following conditions:

existence: a solution exists;

uniqueness: there is only a single solution;

stability: small disturbances in the initial or boundary conditions lead only to small variations of the solution.

The first two conditions confirm a deterministic world view: with given initial and boundary conditions, the solution is known. The third condition for a well-posed problem yields an important restriction. Namely, this condition suggests that a deterministic solution is *in practice* only possible when the solution is not susceptible to small disturbances in the initial and boundary conditions. But why do we consider the notion of 'in practice'? Mathematically speaking, there would be no objection against using exactly known initial and boundary conditions. In that case we speak of mathematically *ideal* initial and boundary conditions, and for such a situation the solution would be completely determined due to the first and second conditions. However, this requires the initial and boundary conditions to be known with *infinite accuracy*; this would of course 'in practice' not be feasible, since the initial and boundary conditions are only known with *finite accuracy*. This is what we refer to as *realistic* initial and boundary conditions. Only for a well-posed problem these imperfections in the initial and boundary conditions fail to significantly affect the solution. In that case, the solution is 'in practice' completely predictable. If, however, the third Hadamard condition is not satisfied, we can expect completely different behavior of our solution when we do not exactly know the initial and boundary conditions, which results in what we perceive as *unpredictability*. In that case, the problem is considered to be 'ill-posed'.

These ideas were first elaborated by the French mathematician Henri Poincaré, who published his work *Méthodes Nouvelles de la Mécanique Céleste* in 1892. In this work he tried to solve the famous so-called *three-body problem*. Newton had already solved the *two-body problem*, where he found the elliptic Kepler-trajectories as the solution, which are considered the hallmark of a completely predictable solution. In other words, the two-body problem is well posed. This is in sharp contrast to the three-body problem. Poincaré found that this problem is not integrable; in short, this means that there is no simple solution in terms of a smooth or differentiable function. He found that the solution had irregular and chaotic characteristics. The solution to the three-body problem thus appears to be fundamentally unpredictable; the problem is *ill posed*. This meant the end of Laplace's orderly world view.

The results of Poincaré were further elaborated in the theory of dynamical systems. Dynamical systems can be imagined best as systems of coupled differential equations, describing the behavior of so-called system variables as a function of time. Often the number of system variables, or degrees of freedom, is kept relatively small. An example of a dynamical system is that of two coupled pendulums, but other examples of dynamical systems can be found in economics and certain biological processes. Dynamical systems are deterministic by definition, and therefore full solutions as a function of time can *in principle* be computed.

For these systems, consisting of sets of regular differential equations, it was proven that, given an initial condition, a single and unique solution exists. Hence, the first and second Hadamard conditions are satisfied. The third condition, however, is not always satisfied. The validity of this condition can only be proven for a limited number of systems, often only the linear ones. On the other hand, for many nonlinear dynamical systems it has been found that the solution is extremely sensitive to small variations of the initial conditions. The solution then becomes unpredictable after a certain amount of time, after which it starts to fluctuate. We cannot predict the magnitude, and often not even the sign of these fluctuations. This is called *deterministic chaos*. It is one of the most fundamental new insights that has dominated the developments in mechanics over the last couple of decades. It should be emphasized here that such 'chaotic' behavior is only anticipated for *nonlinear* dynamical systems.

Let us consolidate these findings to the solution of our flow equations. In doing so, we should note that the preceding findings have only been demonstrated for dynamical systems where the number of degrees of freedom is small. Nonetheless, we expect similar findings for systems that have many degrees of freedom, although this has only been proven in a few cases. Here, we interpret the equations of motion for the flow as a system with many degrees of freedom.

Consider a solution of the Navier–Stokes equations for a given flow problem; realistic initial and boundary conditions are given with finite accuracy. Suppose that all conditions for a well-posed problem are be satisfied, so that the solution for the flow is completely *predictable*. We define this as *laminar flow*.

However, the Navier–Stokes equations are nonlinear, and thus we have to expect that only under very special circumstances it is possible to comply with the conditions for a well-posed problem, especially the third one. In all other cases, the equations of motion and initial and boundary conditions for the flow would be ill posed. The solution is then susceptible to small variations in the initial or boundary conditions. We argued above that in this case the solution eventually becomes completely unpredictable, and this now defines a *turbulent flow*. In short, turbulence is associated with the concept of *deterministic chaos*, as mentioned above. So, turbulence is a completely different *kind* of flow than laminar flow, which would be unaffected by small variations in the initial and boundary conditions.

Now what does unpredictability on the basis of susceptibility to initial and boundary conditions mean? Suppose that we consider two solutions of a turbulent flow with the same *realistic* initial and boundary conditions. This would mean that for both solutions the initial and boundary conditions may be different, but within a finite degree of accuracy or tolerance. Such small differences will always be present in practice; for example, consider differences that are the result of molecular fluctuations. Since the third condition for a well-posed problem is not satisfied, the two solutions will, after a certain moment, begin to diverge completely. These two solutions can be considered as two *realizations* of the turbulent flow. Thus, every solution with the same realistic initial and boundary conditions yields a completely new realization. For turbulent flow with given realistic initial and boundary conditions, we are unable to predict the flow temporal and spatial evolution of the flow variables.

It now becomes plausible to consider the *statistics* of the flow variables, rather than the individual realizations. This is the most widely applied approach to describe turbulent flows. However, we cannot reverse this argument; from the turbulence statistics we will never be able to reconstruct the full course of all realizations. This underlies the so-called *closure problem*, which has remained the central fundamental issue in the theoretical description of turbulence.

In the preceding part of this Introduction, an attempt was made to relate flow turbulence to contemporary dynamical systems theory. It should be mentioned here that this relation is mainly qualitative. Except for the *routes to chaos* that are treated in Sect. 3.4, the reader will look in vain for a more quantitative elaboration in the remainder of this book. Nonetheless, the common understanding is that concepts and results in modern *chaos theory* have aided to gain insight in the onset of turbulence in flows. Some even say that this is the right way forward that will eventually lead us to a full solution of the problem of turbulence; see for example the preface in the latest edition of the famous book by Landau and Lifshitz on fluid mechanics. However, many still question whether chaos theory will provide us with a theory of turbulence that could, for example, predict the behavior of a turbulent flow. Note that chaos theory has prepared us to accept that we may never reach a closed-form solution to the problem of turbulence; the Navier–Stokes equations appear to be fundamentally *non-integrable*.

Separate from the modern mathematical insights mentioned above, the field of turbulent flows has passed through a long-term development. This mainly originated from practical questions and problems. Ignoring older and primarily qualitative considerations, research on turbulence commenced in the 18th and 19th centuries. Its origins can be found in the field of hydraulics, because of the interest in studying turbulent flow through pipes. Most of these investigations were empirical. This is why the works of Osborne Reynolds in 1883 and 1895 are considered as the birth of the theory of turbulence. His name will return multiple times in the ensuing chapters.

In this book we first focus on developments in the dynamics of turbulent flow. In doing so, phenomenological considerations are often invoked, because there is in fact no satisfactory theory of turbulence. This is despite the fact that many famous physicists, for example Werner Heisenberg and Richard Feynman, worked on the problem. Quoting from the *Feynman Lectures on Physics*:

> Finally, there is a physical problem that is common to many fields, that is very old, and that has not been solved. It is not the problem of finding new fundamental particles, but something left over from a long time ago – over a hundred years. Nobody in physics has been able to analyze it mathematically satisfactorily in spite of its importance to the sister sciences. It is the analysis of *circulating* or *turbulent fluids*.

In absence of a comprehensive theory, we have to resort to heuristics now and then. The use of dimensional analysis, linked with an adequate insight into the physical processes at hand, is then most appropriate. This may leave the reader who expects a comprehensible theory of turbulence somewhat dissatisfied. However, it will be shown that with this approach many important and useful results can be obtained.

To conclude this Introduction, we provide an overview of the material that is covered in this book. After a brief introduction to the governing equations of motion in Chap. 2, we start in Chap. 3 with a short treatise on the emergence of turbulence. On the basis of *linear stability analysis* we discuss the circumstances and conditions under which laminar flow becomes unstable and when we can expect turbulence to appear. Also, we briefly address the transition to fully-developed turbulent flow. This is referred to as the *route to chaos*. As mentioned previously, new insights from chaos theory contributed to this particular topic.

In the next chapter we focus on a particular model of turbulence that has an exact solution, that is *Burgers equation*. Equipped with the knowledge from this model, we discuss the phenomenology of turbulence. Important concepts, such as the *macrostructure* and *microstructure* are introduced, where each is characterized by separate scaling law.

In Chap. 5 we derive the equations of motion for the *mean velocity* in a turbulent flow. Here we are confronted for the first time with the *closure problem*. This closure problem is key to the development of turbulence models; in particular, in this book we pay attention to several closure models.

The first closure model is Ludwig Prandtl's *mixing length hypothesis*, which we apply to turbulent channel flow in Chap. 6. This also serves as an example of turbulence in the vicinity of a solid boundary, or so-called *wall turbulence*. For wall turbulence we can distinguish several regions with different scaling laws. The most important of these regions is the so-called *inertial sublayer* that is characterized by a logarithmic velocity profile. Wall turbulence is opposed to so-called *free turbulence*, which can develop without the restrictive influence of a solid boundary. Examples of free turbulence are discussed in Chap. 6.

The energetic aspects of turbulence are discussed in Chap. 7. These are studied using the equations for the kinetic energy for the mean flow and for the fluctuating velocity in the turbulent flow. These equations lead to two basic results:

- Turbulent kinetic energy is produced in the macrostructure and is dissipated in the microstructure by molecular viscosity;
- Production and dissipation of turbulent kinetic energy are, in a first approximation, in local equilibrium.

We then address the question by what mechanism energy is transferred from the macrostructure to the microstructure. For this, we introduce the vorticity equation in Chap. 8. The process of *vortex stretching* appears to be responsible for the resulting energy transfer from the macrostructure to the microstructure through what is called the *energy cascade process*. Moreover, on the basis of a first-order approximation of the equation for the vorticity fluctuations, we find that the microstructure is indeed in local dynamic equilibrium. This implies that the microstructure is fully decoupled from the macrostructure, and also that the microstructure is *isotropic*. The theory of the turbulence microstructure was first formulated by A.N. Kolmogorov around 1940. A second-order balance of the equation for the fluctuating vorticity leads to an important closure model that is very useful in practice: the so-called k-ϵ model. Subsequently, we discuss also a couple of contemporary turbulence models, such as the *second-order closure model* and the *algebraic stress model*.

Next, in Chap. 9 we discuss the *correlation function* and its Fourier transform: the *spectrum*. Here we mainly limit ourselves to two specific topics. The first topic is the so-called one-dimensional spectrum, which can be interpreted as the spectrum of the fluctuations in a flow property measured at a fixed point. The second topic we discuss is the theory of *isotropic turbulence*. As one of the central results of the theory of turbulence we deduce the so-called $-5/3$-law of the *inertial subrange* of the spectrum. It is demonstrated that this result is directly related to the existence of both the macrostructure and the microstructure, which are dynamically decoupled.

Finally, we conclude our description of turbulence in Chap. 10 with a brief discussion of *turbulent diffusion* and particle-laden turbulent flows.

Almost every section includes selected problems. These are intended to illustrate the covered material, but occasionally expand on more advanced topics.

One final comment. For certain, turbulence is not one of the easiest subjects in fluid mechanics. This book is therefore only an introduction and a gateway to wondrous things beyond.

Chapter 2
Equations of Motion

2.1 Incompressible Flow

The equations of motion describing the flow in a fluid are based on the three laws
of conservation of mass, momentum and energy. For a detailed formulation of these
equations, one of the standard works in fluid mechanics should be studied (see literature list). Here we limit ourselves to a short derivation of these equations.

We start with a so-called *fluid element* whose volume is δV. This fluid element
should be large compared to the molecules that make up the fluid (for example, it
should be much larger than the mean free path λ_v of a gas) while simultaneously
it should be small compared to the smallest dimensions of flow we are going to
describe. The position of this fluid element is determined in a right-handed Cartesian
coordinate system (x_1, x_2, x_3). Using such a fluid element, the so-called *continuum
hypothesis* allows us to define a pressure p, a velocity u_i, and a fluid density ρ at
every point x_i and every instant in time t.

The fluid element moves through the medium with a velocity u_i at a given position
of this element. For this reason, the fluid element is sometimes called a *material
particle*. In the trajectory of this fluid element, alterations as a function of time
are indicated by the so-called *material derivative*: D/Dt. In a Cartesian coordinate
system this material derivative is defined as:

$$\frac{D}{Dt} \equiv \frac{\partial}{\partial t} + u_j \frac{\partial}{\partial x_j}. \tag{2.1}$$

The material derivative consists of two terms: $\partial/\partial t$, which is known as the *local
derivative*, and $u_j \partial/\partial x_j$, which is called the *advection term*. This advection term
describes the part of the material derivative due to transport in a velocity field varying
in space. In the literature this contribution to the material derivative is also called
convection. Here, however, we call it advection in order to distinguish it from *thermal*
convection when a non-isothermal flow is dominated by effects due to variation in
density.

© Springer International Publishing Switzerland 2016

F.T.M. Nieuwstadt et al., *Turbulence*, DOI 10.1007/978-3-319-31599-7_2

As in Eq. (2.1) above, we make use of the *Einstein notation*. This means that a repeated index (as in $u_i u_i$ for example) is summed over all directions of the coordinate system: $u_i u_i \equiv u_1 u_1 + u_2 u_2 + u_3 u_3$. An exception is when we use a *Greek index letter* (for example $u_\alpha u_\alpha$; in that case we simply indicate the term *without* applying the summation.

Conservation of mass for the fluid element defined previously can now be described as

$$\frac{D(\rho\, \delta V)}{Dt} = 0. \tag{2.2}$$

This equation can be reduced to

$$\frac{1}{\delta V}\frac{D\delta V}{Dt} = -\frac{1}{\rho}\frac{D\rho}{Dt}. \tag{2.3}$$

When the following approximation holds, we call a flow incompressible:

$$\frac{1}{\rho}\frac{D\rho}{Dt} \approx 0. \tag{2.4}$$

This appears to be the case when the flow velocities are much smaller than the speed of sound a (i.e., $U \equiv |u_i| \ll a$).

In the remainder of this book we limit ourselves to incompressible flow. Conservation of mass then reduces to

$$\frac{1}{\delta V}\frac{D\delta V}{Dt} \equiv \frac{\partial u_i}{\partial x_i} = 0. \tag{2.5}$$

The derivation of this equation can be found in the problems section. The velocity field is thus shown to be *divergence-free* at all points. Eq. (2.5) is also known as the *continuity equation*.

Next we consider the conservation of momentum for our fluid element. According to Newton's second law it follows that:

$$\rho\, \delta V \frac{Du_i}{Dt} = F_i, \tag{2.6}$$

where F_i represents the net force acting on the fluid element. For F_i we can write

$$F_i = \left(\rho g_i + \frac{\partial \sigma_{ij}}{\partial x_j}\right)\delta V. \tag{2.7}$$

The first term of the equation above is called the *volume force*. In this case we equated this volume force to gravitation with gravitational acceleration $g_i = (0, 0, -g)$. The second term on the right-hand side of (2.7) is called the *surface force* with a *surface stress tensor* σ_{ij}. For incompressible flow and for a Newtonian fluid it follows that

$$\sigma_{ij} = -p\delta_{ij} + \mu \left(\frac{\partial u_i}{\partial x_j} + \frac{\partial u_j}{\partial x_i} \right), \tag{2.8}$$

where δ_{ij} represents the *Kronecker-δ* symbol: $\delta_{ij} = 0$ if $i \neq j$, and $\delta_{ij} = 1$ when $i = j$. The surface stress tensor thus consists of two parts. The first term represents an isotropic pressure. The second term, which relates to deformations of the fluid element, is called the *shear stress*. In this term, μ is known as *dynamic viscosity*. This is a material property, and thus only depends on the fluid. In the following we consider a Newtonian fluid and thus take μ as a constant.

We note that μ is often combined with the fluid density to form the *kinematic viscosity* $\nu = \mu/\rho$. Some representative values for ν at standard atmospheric conditions are: $1.5 \times 10^{-5} \, \text{m}^2/\text{s}$ for air, and $1.0 \times 10^{-6} \, \text{m}^2/\text{s}$ for water.

Substituting (2.8) in (2.7) and then in (2.6) leads to the following equation for the conservation of momentum in an incompressible flow:

$$\rho \frac{Du_i}{Dt} \equiv \rho \left(\frac{\partial u_i}{\partial t} + u_j \frac{\partial u_i}{\partial x_j} \right) = \rho g_i - \frac{\partial p}{\partial x_i} + \mu \frac{\partial^2 u_i}{\partial x_j^2}. \tag{2.9}$$

Eqs. (2.9) (for $i = 1, 2, 3$) are better known as the *Navier-Stokes equations*. Together with the continuity Eq. (2.5), they form a system of four equations containing five unknown variables: u_i, p and ρ. We thus have to specify an additional constraint in order to solve the equations. For this we take $\rho = $ constant, which means that the fluid is *homogeneous*. In other words: the density is reduced to a material constant (for air $\rho \approx 1.2 \, \text{kg m}^{-3}$; for water $\rho \approx 1.0 \times 10^3 \, \text{kg m}^{-3}$).

Another further simplification is possible when no free surface is present in our flow. In that case we can absorb the gravity term in (2.9) in the pressure term, which then is referred to as the *modified pressure*. The equations for the conservation of momentum are then reduced to

$$\rho \frac{Du_i}{Dt} \equiv \rho \left(\frac{\partial u_i}{\partial t} + u_j \frac{\partial u_i}{\partial x_j} \right) = -\frac{\partial p}{\partial x_i} + \mu \frac{\partial^2 u_i}{\partial x_j^2}. \tag{2.10}$$

Equations (2.5) and (2.10) form the basis for describing incompressible flow of a homogeneous fluid. We note that initial and boundary conditions have to be specified before a solution of the equations of motion can be found.

2.1.1 Problem

1. Consider a material line segment δL_x at x_i. The segment is oriented parallel to the x-axis. The change in length of this line segment follows from

$$\frac{D\delta L_x}{Dt} = u(x_i + \delta L_x) - u(x_i),$$

in which u represents the x-component of the velocity vector u_i. Apply a Taylor-series expansion to the right-hand side of the equation above. Next, consider the line segments δL_y and δL_z, which are oriented parallel to he y- and z-axes respectively and repeat the exercise. Now prove that

$$\frac{1}{\delta V} \frac{D \delta V}{Dt} = \frac{\partial u_i}{\partial x_i},$$

using $\delta V = \delta L_x \, \delta L_y \, \delta L_z$.

2.2 The Boussinesq Approximation

In Eqs. (2.5) and (2.10) we accounted for the influence of gravity on flow by absorbing it into the pressure term. We argued that this is only possible for a homogeneous fluid (ρ = constant) in the absence of a free surface.

In practice, however, we are often confronted with flows in fluids where the density depends on position and time. These are called *heterogeneous fluids*. An example is the density variation that can occur due to temperature differences. Also a variable composition of the fluid can lead to variations of the density. An example of this would be the salt concentration in the ocean.

The dynamics of heterogeneous fluids are directly affected by gravity, in which case we speak of a *stratified flow*. The equations of motion for these flows are again based on the laws of conservation of mass, momentum and energy.

For the conservation of mass we use again the continuity equation

$$\frac{\partial u_i}{\partial x_i} = 0, \tag{2.11}$$

provided, of course, that the condition $U \ll a$ is satisfied, as mentioned in the previous section. Also, we need to take into account the fact that, in a heterogeneous fluid, the density varies with height due to gravity. This is described by the *hydrostatic law*: $\partial p / \partial z = -\rho g$. We now introduce the *scale height* as a measure of the distance over which the density varies due to gravity. This scale height H is defined as

$$H = \left(-\frac{1}{p} \frac{\partial p}{\partial z} \right)^{-1} = \frac{p_0}{\rho g}. \tag{2.12}$$

where p_0 is a reference pressure (for example the pressure at the surface for an atmospheric flow). It follows that (2.11) is only valid when $L \ll H$, where L is representative of the extent of the vertical motion in the flow.

Applying the conservation of momentum leads again to (2.9). However, the density ρ in (2.9) should now be seen as an unknown variable, which we have to compute as a function of position and time. This means that we need an additional equation. For this, we use a so-called *equation of state*, which reads, in general form

$$\rho = \rho(p, \theta), \tag{2.13}$$

where θ represents temperature. We first limit ourselves to liquids for which the *compressibility modulus* $(\partial\rho/\partial p)$ is negligible. This means that ρ is only a function of temperature. Now we just need an additional equation describing the temperature, and for this we apply the equation for the *conservation of energy* to a fluid element. For a full discussion on the energy equation, one should consult one of the standard works on thermodynamics or fluid mechanics. It follows that, for a liquid, the energy equation can be approximated by the following equation for the temperature:

$$\frac{\partial\theta}{\partial t} + u_j\frac{\partial\theta}{\partial x_j} = \alpha\frac{\partial^2\theta}{\partial x_j^2}, \tag{2.14}$$

where α is the *thermal diffusivity coefficient*. For air the value for α is $\sim 2.0 \times 10^{-5}$ $m^2 s^{-1}$, and for water $\sim 1.4 \times 10^{-7}$ $m^2 s^{-1}$. The ratio between the material constants ν and α is referred to as the *Prandtl number*

$$Pr = \frac{\nu}{\alpha}. \tag{2.15}$$

For air we find $Pr \approx 0.7$ and for water $Pr \approx 8$.

The set of Eqs. (2.11), (2.9) and (2.14) may now form a closed system, but is still too complicated to solve. This is why we need to simplify them by applying the so-called *Boussinesq approximation*. The first step of this approximation is the definition of a *reference state*: p_0, ρ_0 and T_0. This reference state has to obey the equations of motion for the fluid at rest

$$\theta_0 \equiv T_0 = \text{constant} \quad \text{and} \quad \frac{\partial p_0}{\partial z} = -\rho_0 g. \tag{2.16}$$

We now consider a flow with velocity u_i, pressure $p_0 + p$, density $\rho_0 + \rho$, and temperature $T_0 + \theta$. We impose the conditions that $p/p_0 \ll 1$, $\rho/\rho_0 \ll 1$, and $\theta/T_0 \ll 1$. In other words, p, ρ and θ are small disturbances with respect to the reference state. Substitution in (2.9) leads to

$$(\rho_0 + \rho)\frac{Du_i}{Dt} = (\rho_0 + \rho)g_i - \frac{\partial}{\partial x_i}(p_0 + p) + \mu\frac{\partial^2 u_i}{\partial x_j^2}. \tag{2.17}$$

After multiplying this equation by $1/\rho_0$, and after substitution of the equation of motion (2.16) for p_0, we find

$$\frac{Du_i}{Dt} = \frac{\rho}{\rho_0}g_i - \frac{1}{\rho_0}\frac{\partial p}{\partial x_i} + \frac{\mu}{\rho_0}\frac{\partial^2 u_i}{\partial x_j^2}. \tag{2.18}$$

Here we used the condition $\rho/\rho_0 \ll 1$ for the left-hand side of (2.17), so that $(\rho_0 + \rho)Du_i/Dt \approx \rho_0 Du_i/Dt$. However, this simplification is not applied to the term $(\rho_0 + \rho)g_i$. The physical background for this is that, for all flows considered here, $g \gg |Du_i/Dt|$.

Next, we apply a linearization of the equation of state (2.13) around θ_0. Now we have

$$\frac{\rho}{\rho_0} = -\beta\frac{\theta}{T_0}, \tag{2.19}$$

where β is the *volumetric expansion coefficient*:

$$\beta = -\frac{T_0}{\rho_0}\frac{\partial \rho}{\partial \theta}.$$

One has to pay careful attention here! We cannot apply the linearization to the velocity term, that is, $Du_i/Dt \approx \partial u_i/\partial t$. The reason for this is that the velocity field is zero for the reference state. Hence, the velocity is never small.

The result of these calculations is a set of three equations, which are referred to as the *Boussinesq equations*

$$\frac{\partial u_i}{\partial x_i} = 0, \tag{2.20}$$

$$\frac{\partial u_i}{\partial t} + u_j\frac{\partial u_i}{\partial x_j} = -\beta\frac{\theta}{T_0}g_i - \frac{1}{\rho_0}\frac{\partial p}{\partial x_i} + \nu\frac{\partial^2 u_i}{\partial x_j^2}, \tag{2.21}$$

$$\frac{\partial \theta}{\partial t} + u_j\frac{\partial \theta}{\partial x_j} = \alpha\frac{\partial^2 \theta}{\partial x_j^2}. \tag{2.22}$$

The set of equations above has, in principle, been derived for liquids. However, the energy Eq. (2.14) is also applicable to gases, provided that θ is interpreted as the so-called *potential temperature*. In this way the compressibility of a gas with height is accounted for. The potential temperature is defined as the temperature of a fluid element with pressure p and temperature T when it is brought to a standard pressure p_* via an isentropic process. For an ideal gas it follows that

$$\theta = T\left(\frac{p_*}{p}\right)^{\frac{\kappa-1}{\kappa}}, \tag{2.23}$$

where $\kappa = c_p/c_v$ is the ratio between the *specific heat at constant pressure* c_p and the *specific heat at constant volume* c_v. For air, $c_p = 1005\,\mathrm{J\,kg^{-1}\,K^{-1}}$ and $R(= c_p - c_v) = 287\,\mathrm{J\,kg^{-1}\,K^{-1}}$, so that $\kappa \approx 1.4$.

It should be mentioned here that Eq. (2.23) is commonly known as the *Poisson equation* for an isentropic process, i.e. the entropy S of the fluid element remains constant during the process. It follows from this definition of potential temperature that $\theta \sim S$. The background for this is that the energy equation, using the concept of entropy, can be written in its most general form as

$$\frac{DS}{Dt} = Q_S$$

where Q_S represents all processes that increase entropy, such as molecular conduction. Simplification of this equation forms the fundamental basis for Eq. (2.14).

We can now calculate the difference between the normal temperature T and the potential temperature θ. We take the atmosphere as an example, for which $\kappa = 1.4$. The standard pressure p_* is taken to be equal to 1000 mbar (or 10^5 Pa), which is approximately the pressure at ground level. If p does not vary too much from p_*, in other words $T \approx \theta$ (that is, we limit ourselves to the lower layer of the atmosphere, which is the so-called *atmospheric boundary layer*), then substitution of (2.23) in the hydrostatic law (2.16) yields for p

$$\frac{\partial \theta}{\partial z} = \frac{\partial T}{\partial z} + \gamma_d, \quad \text{with:} \quad \gamma_d = \frac{g}{c_p}. \tag{2.24}$$

Here γ_d is the so-called *adiabatic temperature gradient*. For our atmosphere this is $0.01°C\,m^{-1}$. This means that we obtain the potential temperature by multiplying the temperature in the atmosphere with $\gamma_d z$ (where z is measured from the surface of the earth).

It is clear that the correction of the potential temperature matters only to the atmosphere; this term is therefore often seen in the meteorological literature. In a laboratory, z is virtually negligible with respect to the atmospheric scale height. This means that flows of liquids and gases are indistinguishable in the laboratory. That is why, from now on, we use the term 'temperature' for θ, rather than 'potential temperature' (although one should be aware that in the atmosphere a correction should be applied to the term).

In addition, we also limit ourselves to *ideal gases*, that is: $p = \rho R T$. For such gases the volumetric expansion coefficient β is unity ($\beta = 1$).

2.2.1 Problems

1. Compute the scale height of an isothermal atmosphere (meaning that T is constant with increasing height). We consider air to be an ideal gas: $p = \rho R T$, where R is the gas constant, which has a value for air of $R = 287.04\ m^2\ s^{-2}\ K^{-1}$. Show that turbulence occurring in the atmospheric boundary layer (the lower few kilometers of the atmosphere) can be considered as an incompressible flow, while

compressibility is not negligible for turbulence related to thunderstorms (which are spread out over the lower 10 Km of the atmosphere).

2. Consider the atmospheric boundary layer at rest with an average temperature gradient dT_0/dz = constant as the initial state (the average potential temperature gradient is then equal to $dT_0/dz + \gamma_d$). We consider the vertical motion of material air particles. Show that for small disturbances, neglecting molecular effects, it follows that for the vertical position z_p of an air particle

$$\frac{d^2 z_p}{dt^2} + N^2 z_p = 0,$$

where:

$$N^2 = \frac{g}{T_0}\left(\frac{dT_0}{dz} + \gamma_d\right) \equiv \frac{g}{T_0}\left(\frac{d\Theta_0}{dz}\right)$$

is called the *Brunt–Väisälä frequency*.

At $t = 0$ we shift the position of the particle by $z_p(0)$ relative to its initial position. Using the equation above it follows that for

(a) $dT_0/dz = -\gamma_d$ (or: $d\Theta_0/dz = 0$) the particle will not move. The atmosphere is called *neutral*.

(b) $dT_0/dz > -\gamma_d$ (or: $d\Theta_0/dz > 0$) the particle will start oscillating with frequency N. The atmosphere is called *stable*.

(c) $dT_0/dz < -\gamma_d$ (or: $d\Theta_0/dz < 0$) the particle is unstable. The atmosphere is called *convective*.

Fig. 2.1 Plume from a smokestack in a stable atmosphere

Fig. 2.2 Definition of
coordinate systems in the
laboratory and in the
atmosphere

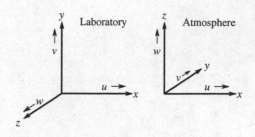

Fig. 2.1 shows a plume from a smokestack on a cold winter day; despite a strong
wind there is little mixing due to the strong stratification of the atmosphere. The
clear sky allows a strong radiation of heat, so that the temperature near the surface
is lower than higher up in the atmosphere (i.e., $d\Theta_0/dz > 0$).

2.3 Coordinate System

As mentioned in Sect. 2.1, we primarily use a *Cartesian coordinate system*: (x_1, x_2, x_3)
or (x, y, z). However, two conventions for the vertical axis can be used.

In the so-called *laboratory coordinate system* the y-axis is taken vertical. However,
for flows in the atmosphere and in the ocean, most often the z-axis is chosen as the
vertical axis. Both coordinate systems, illustrated in Fig. 2.2, are used throughout
this book.

Chapter 3
Stability and Transition

3.1 Stability Analysis

Suppose we acquired a solution to the Navier–Stokes and Boussinesq equations for a certain flow geometry. Of course the next question is whether this solution can be realized in practice. With this we arrive at a subject we discussed previously in Chap. 1, i.e. whether the problem is *well posed* or *ill posed*. In that chapter it was stated that for given ('*mathematically ideal*') initial and boundary conditions, it is *in principle* always possible to find a solution of the Navier–Stokes equations. However, only when the problem is well posed, small perturbations in the initial and boundary conditions will remain small, and thus only then a solution can be realized; in that case the flow is called '*laminar*'. In the case of an ill-posed problem, small disturbances or perturbations lead instead to finite perturbations in the solution, capable of growing rapidly in a finite amount of time. Such solutions are not physically feasible; as a result, the flow becomes unpredictable. This behavior, which we call *deterministic chaos*, is associated with a turbulent state of the flow.

It seems obvious to start studying the circumstances for which the solutions to the Navier–Stokes equations are physically feasible, that is, when the flow is laminar. In other words, under what circumstances will the influence of the perturbations in the initial and boundary conditions remain small? In order to find the answer to this question, we use *stability analysis*.

Stability analysis works as follows: very small, i.e. *infinitesimal*, perturbations are superimposed on a certain solution of the Navier–Stokes equations, which we call the *base flow*. Next, with the use of the equations of motion, it is calculated how these perturbations behave as a function of time. When the perturbations remain small, we call the solution *stable*, and the flow is in a *laminar* flow state. When the perturbations grow as a function of time, the solution is called *unstable*, and we expect that the solution leads to a *turbulent* flow state.

In this analysis, we make use of the fact that the applied perturbations are initially small (i.e., infinitesimal). In that case the quadratic and higher-order terms in the equations of motion for the perturbations are negligible. Since the equations for the

© Springer International Publishing Switzerland 2016
F.T.M. Nieuwstadt et al., *Turbulence*, DOI 10.1007/978-3-319-31599-7_3

perturbations remain linear, which allows us to avoid all the problems associated with the nonlinearity of the complete equations of motion, we therefore speak of *linear* stability analysis.

It should be clear that the linear stability theory is only a first step on the way to investigating the emergence of turbulence. The theory can certainly not be used to calculate the development into a fully turbulent flow. After all, in fully turbulent flow, the perturbations can no longer considered to be small, and thus the higher-order nonlinear terms can not be neglected any more. Indeed, in one of the later chapters we see that the nonlinear terms in particular determine the essence of turbulent flow.

However, linear stability analysis remains a useful instrument, because it yields the conditions for which solutions of the Navier–Stokes equations become unstable. These conditions might just determine when a flow becomes turbulent, and thus provides us with some information on how turbulence is generated. We illustrate this in the remainder of this chapter using some specific examples, and we conclude by discussing how an unstable flow becomes fully turbulent.

As a guide for the following discussions, we first list the steps that make up a linear stability analysis:

1. A base flow is defined that represents a *particular solution* to the equations of motion. In general, we take for this the stationary solution, which means that we consider flow variables that are not a function of time.
2. On this base flow we superimpose a small non-stationary, often wave-like, perturbation.
3. The combination of the perturbation and the basic flow is substituted into the equations of motion. The higher-order perturbation terms are ignored (under the assumption that the perturbations are essentially small), and a set of linear differential equations for the perturbations remains.
4. This set of linear differential equations is *homogeneous*; this also holds for the boundary conditions we impose on the perturbations. The set of linear differential equations can be interpreted as a generalized *eigenvalue problem*.
5. A nontrivial solution of this eigenvalue problem is only possible under certain mathematical conditions. These conditions determine the development of the perturbation as function of time.
6. In case the perturbation *grows* as a function of time, the flow is called *unstable*; if the perturbation *decays* as a function of time the flow is called *stable*. If the perturbation is independent of time, the flow is called *neutrally stable*.

3.2 Kelvin–Helmholtz Instability

Consider a base flow as illustrated in Fig. 3.1. At $y = 0$ there is an interface between the two fluid regions. Above the interface the velocity is $-\frac{1}{2}U$, below the interface the velocity is $+\frac{1}{2}U$. At the interface, there is thus a *discontinuity* in the tangential velocity. Furthermore, the static pressure P is assumed to be the same in the whole

Fig. 3.1 Flow geometry for
the Kelvin–Helmholtz
instability

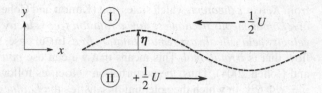

domain. Hence, this velocity profile is the solution of the frictionless equations of motion (2.10) in which the density variations are neglected.

We now determine the stability of this flow, neglecting the effect of the viscosity. This approach is referred to as the *inviscid* stability analysis. In accordance with the procedure described previously, we apply small perturbation η_s to the interface, combined with small perturbations u, v and p to the velocity and pressure. From here on we denote the base flow with uppercase symbols, while lowercase symbols are used for the perturbations.

The first question that has to be answered is: what type of perturbation should be considered? The answer depends on the geometry of the base flow, which in this case is defined for $-\infty < x < \infty$. Furthermore, given that U is independent of x, we can use *Fourier's theorem*. This theorem states that every perturbation (provided that it remains bounded for $x \rightarrow \pm\infty$) can be represented by a Fourier series. It is thus sufficient to limit our analysis to a Fourier series, that is to wave-like perturbations, given by:

$$\eta_s(x, t) = A \, e^{i(kx-\omega t)}, \tag{3.1}$$

where A denotes the amplitude of the wave. In principle, A is a complex number. However, without explicitly repeating this, we consider here only the real part of the expression in (3.1). The wavenumber k relates to the wavelength λ of the perturbation according to $\lambda = 2\pi/k$. In order to apply Fourier's theorem as mentioned above, we have to demand that the wavenumber k is real; if k would be complex, then the perturbation (3.1) would not be bounded for either $x \rightarrow \infty$ or $x \rightarrow -\infty$. We take ω in (3.1) to be complex:

$$\omega = \omega_r + i\,\omega_i. \tag{3.2}$$

The real part ω_r is the *angular frequency*, which relates to the *period* T of the wave according to $T = 2\pi/\omega_r$. Using this, we can interpret (3.1) as a *traveling wave* that propagates with a *phase velocity* c that equals:

$$c = \omega_r/k. \tag{3.3}$$

The imaginary part of ω gives the variation of the amplitude as a function of time. If $\omega_i > 0$, the amplitude grows exponentially, in which case we call the flow *unstable*. In other words, the goal of our stability analysis is to determine the value of ω_i.

We first calculate the velocity components u and v. One of the starting points is that in both regions the flow is *irrotational* (i.e., free of rotation). This directly follows from *Kelvin's theorem*, which states that (Kundu and Cohen 2004): *if at any moment in a homogeneous frictionless flow a rotation free velocity field can be observed, the velocity field will always remain rotation free*. In our case, the starting point is a base flow that is *irrotational*. This means that we can use *potential flow theory* (Kundu and Cohen 2004). Thus, the perturbation velocities follow from: $u = \partial\Phi/\partial x$, and: $v = \partial\Phi/\partial y$, in which the potential Φ satisfies the *Laplace equation*:

$$\Delta\Phi \equiv \frac{\partial^2 \Phi}{\partial x^2} + \frac{\partial^2 \Phi}{\partial y^2} = 0. \tag{3.4}$$

Depending on our choice for the perturbation in η_s, and in accordance with (3.1), we choose:

$$\Phi = F(y)e^{i(kx-\omega t)}. \tag{3.5}$$

Substitution in (3.4) gives:

$$\frac{d^2 F}{dy^2} - k^2 F = 0,$$

which has a solution:

$$F = C_1 e^{-ky} + C_2 e^{ky},$$

in which the integration constants C_1 and C_2 follow from the boundary conditions. In this case these conditions are that the perturbation vanishes for $y \to \pm\infty$. This leads to: $C_1 = 0$ in region II and $C_2 = 0$ in region I, so that

$$\begin{aligned}
\Phi_{\mathrm{I}} &= B_{\mathrm{I}} e^{-ky} e^{i(kx-\omega t)}, \\
\Phi_{\mathrm{II}} &= B_{\mathrm{II}} e^{ky} e^{i(kx-\omega t)}.
\end{aligned} \tag{3.6}$$

Here we replaced C_1 and C_2 by B_{I} and B_{II} respectively, which can be determined by applying the boundary conditions at the interface.

At the interface, two types of boundary conditions can be formulated: a *kinematic* one and a *dynamic* one. The kinematic boundary condition implies that the interface can be regarded as a *material surface*. This implies that η_s should obey

$$\frac{D\eta_s}{Dt} \equiv v(y = \eta_s) = \frac{\partial\Phi}{\partial y}. \tag{3.7}$$

Applying this equation to the interface $y = \eta_s$ in regions I and II yields:

$$\begin{aligned}
\text{in region I}(y \downarrow \eta_s): \quad &\frac{D\eta_s}{Dt} \equiv \frac{\partial\eta_s}{\partial t} + \left(u_{\mathrm{I}} - \frac{1}{2}U\right)\frac{\partial\eta_s}{\partial x} = \frac{\partial\Phi_{\mathrm{I}}}{\partial y}, \\
\text{in region II}(y \uparrow \eta_s): \quad &\frac{D\eta_s}{Dt} \equiv \frac{\partial\eta_s}{\partial t} + \left(u_{\mathrm{II}} + \frac{1}{2}U\right)\frac{\partial\eta_s}{\partial x} = \frac{\partial\Phi_{\mathrm{II}}}{\partial y}.
\end{aligned} \tag{3.8}$$

We retain only the linear terms, so that the term $u\,\partial\eta_s/\partial x$ disappears, and we apply the two conditions (3.8) to $y = 0$ instead of $y = \eta_s$, given that η_s is small. Using (3.1) and (3.6) we find the following relations for B_I and B_{II}:

$$-kB_I = iA\left(-\omega - \frac{1}{2}Uk\right),$$

$$kB_{II} = iA\left(-\omega + \frac{1}{2}Uk\right). \tag{3.9}$$

Next we consider the dynamic boundary condition. It states that all forces (in our case only the pressure) should be continuous across the interface. For a potential flow the pressure can be calculated from *Bernoulli's law*. For both flow regions, this reads:

$$\rho_I\frac{\partial\Phi_I}{\partial t} + \frac{1}{2}\rho_I\left\{\left(-\frac{1}{2}U + u_I\right)^2 + v_I^2\right\} + (P + p_I) + \rho_I g\eta_s = C_I,$$

$$\rho_{II}\frac{\partial\Phi_{II}}{\partial t} + \frac{1}{2}\rho_{II}\left\{\left(\frac{1}{2}U + u_{II}\right)^2 + v_{II}^2\right\} + (P + p_{II}) + \rho_{II} g\eta_s = C_{II}, \tag{3.10}$$

where C_I and C_{II} are, in principle, two different constants.

The equations in (3.10) are formulated for the general case that the density ρ_I in region I is different from the density ρ_{II} in region II. For reasons of simplicity we assume that $\rho = \rho_I = \rho_{II}$. Consequently, the gravity term $g\eta_s$ in (3.10) has no further influence on the solution. (See Problem 1 below for an analysis that includes two fluids with different densities.) The remaining equations should hold for $y \to \pm\infty$, where u, v and p vanish. For the constants C_I and C_{II} this implies that: $C_I = C_{II} = P + \frac{1}{2}\rho(\frac{1}{2}U)^2$. After applying (3.10) above and below the interface, and after linearization of the expressions, it follows:

$$\frac{\partial\Phi_I}{\partial t} - \frac{1}{2}U\frac{\partial\Phi_I}{\partial x} + \frac{p_I}{\rho} + g\eta_s = 0,$$

$$\frac{\partial\Phi_{II}}{\partial t} + \frac{1}{2}U\frac{\partial\Phi_{II}}{\partial x} + \frac{p_{II}}{\rho} + g\eta_s = 0. \tag{3.11}$$

The dynamic boundary condition requires that: $p_I = p_{II}$, and after substituting the expressions for Φ in (3.11) this leads to:

$$iB_I\left(-\frac{1}{2}Uk - \omega\right) = iB_{II}\left(\frac{1}{2}Uk - \omega\right). \tag{3.12}$$

Equations (3.9) and (3.12) form a set of three linear algebraic equations for the variables A, B_I and B_{II}. This set only has a non-trivial solution (that is, a solution that is not identical zero) when the determinant of the linear system equals zero. This

Fig. 3.2 Example of a
Kelvin–Helmholtz instability
from a numerical simulation
of a jet issuing with a
uniform profile through a
round hole in a spherical
volume (Vuorinen et al.
2011) See also Fig. 3.3

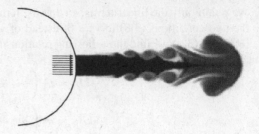

condition can be used to calculate ω. In mathematical terms: we can interpret (3.9) and (3.12) as a generalized *eigenvalue problem*, where ω is the eigenvalue.

The determinant of the set of Eqs. (3.9) and (3.12) reads:

$$\left(i\omega + \frac{1}{2}Uik\right)k\left(\frac{1}{2}iUk + i\omega\right) + k\left(i\omega - \frac{1}{2}iUk\right)\left(-\frac{1}{2}iUk + i\omega\right) = 0, \quad (3.13)$$

which reduces to:

$$\omega^2 = -\frac{1}{4}U^2k^2, \quad \text{or:} \quad \omega = \pm\frac{1}{2}i\,Uk. \quad (3.14)$$

We observe that ω is indeed complex, and that there *always* exists a solution with $\omega_i > 0$. We have to conclude that the flow is *always* unstable for *every* perturbation. This instability, i.e. the instability related to the base flow depicted in Fig. 3.1, is known as the *Kelvin–Helmholtz instability*.

As a result of the nonlinear terms, the growing wave deforms. The final result is a cylindrical vortex structure with its axis positioned normal to the direction of the base flow, as illustrated in Fig. 3.2. These vortices naturally occur in flows with (asymptotic) discontinuities in the velocity. Examples of these are mixing layers and the edges of free jets. This last example is illustrated in Fig. 3.3. Also, Kelvin–Helmholtz waves are sometimes visible atop cloud layers under specific atmospheric conditions (see Fig. 3.4).

A slightly different situation occurs when, aside from a velocity discontinuity, also a jump in the density occurs, i.e. when $\rho_\mathrm{I} \neq \rho_\mathrm{II}$. The only difference compared to the analysis above is that the term $g\eta_s$ contributes to the Bernoulli equation in (3.10). The resulting determinant of the set of linear algebraic equations is then different, and the result now reads:

$$\omega^2 = -\frac{1}{4}U^2k^2 + \frac{1}{2}g\frac{\Delta\rho}{\overline{\rho}}k\left(1 + \frac{2U\omega}{g}\right),$$

where: $\Delta\rho = \rho_\mathrm{II} - \rho_\mathrm{I}$, and: $\overline{\rho} = \frac{1}{2}(\rho_\mathrm{II} + \rho_\mathrm{I})$.

In most cases we can take: $2U\omega \ll g$. (This is a condition identical to the Boussinesq approximation, which states that the acceleration of the velocity field is negligible compared to gravity.) In that case:

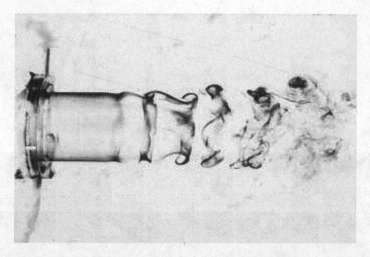

Fig. 3.3 A laminar free jet turning into a turbulent free jet due to a Kelvin–Helmholtz instability. The instability leads here to vortex rings that merge before the complete flow becomes turbulent. See also Fig. 6.18

Fig. 3.4 Kelvin–Helmholtz instability in the atmosphere

$$\omega^2 = -\frac{1}{4}U^2 k^2 + \frac{1}{2}g\frac{\Delta\rho}{\overline{\rho}}k.$$

We find that the flow is *always* unstable when: $\Delta\rho < 0$ (even when $U = 0$). This is obvious, because in this case the heavier fluid is on top of the lighter fluid. For $\Delta\rho > 0$, that is, the heavier fluid is found beneath the lighter one, the flow will be unstable when:

$$k > k_* \equiv \frac{\Delta\rho\, g}{\frac{1}{2}\overline{\rho}U^2}.$$

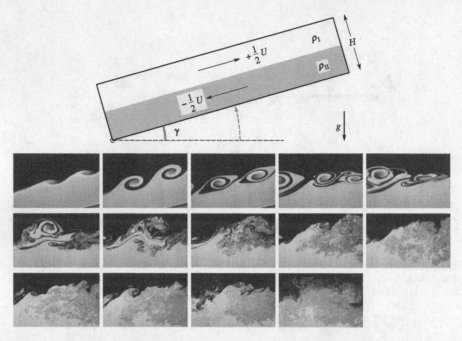

Fig. 3.5 Visualization of the roll-up, pairing, and breakdown of the Kelvin–Helmholtz vortices in a weakly stratified two-dimensional mixing layer with $Re = 2200$ and $Ri = 0.022$ in a tilted tank with two fluids with different density ($\Delta\rho = 0.01$, tilt angle $\gamma = 16°$; see also Problem 1 below). The image width corresponds to about 12 cm. The duration of the image sequence is about $7.7\,\tau$, with $\tau = \frac{1}{2}\lambda/\Delta U$. From: Atsavapranee and Gharib (1997)

This means that only short waves are unstable. For long waves satisfying $k < k_*$, it holds that: $\omega_i \equiv 0$. Under such a condition the flow is called *stable*. The resulting flow consists of waves propagating along the interface. Figure 3.5 shows the development of a planar mixing layer with a small density difference in a channel with a finite height (see also Problem 1 below); note how small waves are most unstable and how the mixing layer rapidly transits to a turbulent flow state.

Problems

1. Consider a flow in a two-dimensional channel with a width of H. In the upper half of the channel the speed is $-\frac{1}{2}U$ and the density is ρ_I, while in the lower half of the channel the speed is $+\frac{1}{2}U$ and the density is ρ_{II} (see Fig. 3.6). Consider small perturbations at the interface of both fluids. Derive that, using the Boussinesq approximation, the condition for stability reads:

$$\left(\frac{\omega H}{U}\right)^2 = -\frac{1}{4}(kH)^2 + \frac{1}{2}F^{-1}kH \tanh\left(\frac{1}{2}kH\right),$$

Fig. 3.6 Flow geometry for Problem 1

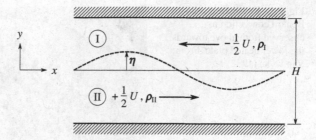

where:

$$F = \frac{\bar{\rho} U^2}{g \Delta \rho H}$$

represents a *Froude number*, with: $\Delta\rho = \rho_{\mathrm{II}} - \rho_{\mathrm{I}}$, and $\bar{\rho} = \frac{1}{2}(\rho_{\mathrm{II}} + \rho_{\mathrm{I}})$. Show that the flow is unstable for *every* perturbation when $F < 0$ or $F > 1$. Give an interpretation for the Froude number.

2. Consider a liquid medium between two horizontal infinite plates at a distance of H from each other. The origin of the coordinate system lies on the lower plate and z represents the vertical coordinate. The lower plate has a temperature T_1 and the upper plate has temperature T_2, with $T_1 > T_2$. The velocity is zero everywhere in the initial state, and the temperature is initially given by the equation

$$T = T_1 - \frac{z}{H}\Delta T,$$

with: $\Delta T = (T_1 - T_2)$.

Superimpose small perturbations for the temperature θ and the velocity u_i on the initial state. The resulting total flow (i.e., base flow plus small perturbations) satisfies the Boussinesq equations (2.20–2.22). Solve the linear perturbation problem with the boundary conditions: $\theta = 0$, $w = 0$ and $\partial u/\partial z = \partial v/\partial z = 0$ at $z = 0$ and $z = H$. These last two conditions mean that the flow along the two horizontal plates is frictionless.

Show that the base flow becomes unstable when the *Rayleigh number*

$$Ra \equiv \frac{g H^3 \Delta}{T_0 \nu \kappa} = \frac{27}{4}\pi^4,$$

and that the most unstable wavelength λ equals:

$$\lambda = \frac{4}{\sqrt{2}}H.$$

This instability is called *Bénard convection*. This instability leads to a flow consisting of so-called *convection cells* that are illustrated in Fig. 3.7.

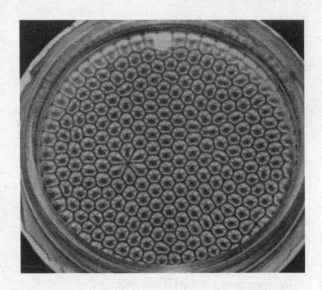

Fig. 3.7 Convection cells in Bénard convection as observed from above towards the lower heated plate. From: Van Dyke (1982)

3.3   Stability of a One-Dimensional Flow

We saw in the previous section that a flow with a discontinuity in the velocity profile is always unstable. In this section we focus on the linear stability analysis of a flow with a *continuous* velocity profile.

As the base flow we consider a so-called *one-dimensional flow*. This means that the velocity vector has only one non-zero component, for which choose the x-component, so that: $u_i = (U, 0, 0)$. In addition, we assume that all flow variables, such as the velocity U, the pressure P and the temperature Θ are functions of one coordinate only, for we take the z-direction. This flow geometry is illustrated in Fig. 3.8. An example of one-dimensional flow is the *channel flow*, which we discuss in more detail in Chap. 6. Another example of a one-dimensional flow, as depicted in Fig. 3.8, is the boundary layer over a flat plate, provided that the (slow) development of the boundary layer in the x-direction is neglected.

A flow with $U = U(z)$, $P = P(z)$ and $\Theta = \Theta(z)$ is a solution to the frictionless, or inviscid, Boussinesq equations when it satisfies:

$$\frac{1}{\rho_0}\frac{\partial P}{\partial z} - \frac{g}{T_0}\Theta = 0. \tag{3.15}$$

According to the procedure for linear stability analysis, we superimpose small perturbations w', p', and θ' on the base flow. Here we limit ourselves to *two-dimensional perturbations*, that is u', w', p' and θ' are functions of x and z only. Actually, according to *Squire's theorem* it can be demonstrated that the critical Reynolds number above which perturbations become unstable (see Fig. 3.10) is lower for two-dimensional perturbations than for three-dimensional perturbations.

Fig. 3.8 Definition of a
one-dimensional flow

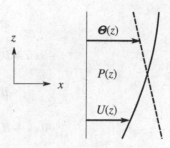

The total velocity, pressure and temperature, i.e. $U + u'$, w', $P + p'$ and $\Theta + \theta'$, should obey the complete equations of motion. After substitution of the perturbations in the equations of motion and applying (3.15), the following set of *linear partial differential equations* can be obtained, where quadratic and higher-order terms have been ignored:

$$\frac{\partial u'}{\partial x} + \frac{\partial w'}{\partial z} = 0,$$

$$\frac{\partial u'}{\partial t} + U\frac{\partial u'}{\partial x} + w'\frac{\partial U}{\partial z} = -\frac{1}{\rho_0}\frac{\partial p'}{\partial x},$$

$$\frac{\partial w'}{\partial t} + U\frac{\partial w'}{\partial x} = -\frac{1}{\rho_0}\frac{\partial p'}{\partial z} + \frac{g}{T_0}\theta', \tag{3.16}$$

$$\frac{\partial \theta'}{\partial t} + U\frac{\partial \theta'}{\partial x} + w'\frac{\partial \Theta}{\partial z} = 0.$$

Based on the same line of reasoning as applied in the previous section, we limit ourselves to wavelike perturbations:

$$u_i' = \hat{u}_i(z)e^{ik(x-ct)}$$

$$\theta_i' = \hat{\theta}_i(z)e^{ik(x-ct)} \tag{3.17}$$

$$p' = \hat{p}_i(z)e^{ik(x-ct)}$$

where the wavenumber k is real by definition. The *phase velocity c* can be complex: $c = c_r + ic_i$. The real part c_r represents the phase velocity of the wave; the imaginary part c_i is the exponent of the growth rate, where: $c_i > 0$ is an indicator for an unstable perturbation. The terms $\hat{u}_i$, $\hat{\theta}$ and $\hat{p}$ represent complex amplitudes that are all functions of z. The expressions (3.17) are complex, but we always consider the real part of the expression, unless indicated otherwise. Substitution of the perturbations (3.17) in (3.16) gives four ordinary differential equations for the complex perturbation amplitudes as a function of z:

$$ik\hat{u} + \frac{d\hat{w}}{dz} = 0,$$

$$-ikc\hat{u} + ik\hat{u}U + \hat{w}\frac{dU}{dz} = -ik\frac{\hat{p}}{\rho_0},$$

$$-ikc\hat{w} + ik\hat{w}U = -\frac{1}{\rho_0}\frac{d\hat{p}}{dz} + \frac{g}{T_0}\hat{\theta},$$

$$-ikc\hat{\theta} + ik\hat{\theta}U + \hat{w}\frac{d\Theta}{dz} = 0.$$

(3.18)

We eliminate $\hat{u}$ using the first of these equations, and subsequently we eliminate $\hat{p}$ from the second and third equations. From the resulting equations we then eliminate $\hat{\theta}$, which results in a single equation for $\hat{w}$, which we call the *Taylor–Goldstein equation*:

$$(U - c)^2\left\{\frac{d^2\hat{w}}{dz^2} - k^2\hat{w}\right\} - \left\{(U - c)\frac{d^2U}{dz^2} - \frac{g}{T_0}\frac{d\Theta}{dz}\right\}\hat{w} = 0. \qquad (3.19)$$

Equation (3.19) is a homogeneous differential equation with homogeneous boundary conditions: $\hat{w} = 0$, or: $d\hat{w}/dz = 0$. A non-trivial solution, that is a solution: $\hat{w} \neq 0$, is only possible under specific conditions. Thus, again we arrive at a generalized eigenvalue problem, and only for certain values of c (i.e., for certain eigenvalues) a non-trivial solution exists. This gives the condition under which the perturbation will either grow or decay as a function of time.

For the Taylor–Goldstein equation no fully closed solution is known. That is why we simply limit ourselves to discuss some examples for which a solution to Eq. (3.19) is available. As our first example we take a base flow with $U = 0$ and $d\Theta/dz = $ *constant*. The Taylor–Goldstein equation then reduces to:

$$\frac{d^2\hat{w}}{dz^2} - \left\{k^2 - \frac{1}{c^2}\frac{g}{T_0}\frac{d\Theta}{dz}\right\}\hat{w} = 0. \qquad (3.20)$$

The solution to this equation is: $\hat{w} = B\exp(imz)$, where m can be interpreted as a *vertical wavenumber*. The value of c follows from the condition that $\hat{w}$ is a non-trivial solution. Substitution of this solution for $\hat{w}$ in (3.20) yields:

$$c^2 = \frac{\frac{g}{T_0}\frac{d\Theta}{dz}}{m^2 + k^2}. \qquad (3.21)$$

This expression, describing the relationship between c and the wavenumbers k and m, is called a *dispersion relation*.

We can distinguish two cases:

1. $d\Theta/dz > 0$

   The solution for c is real. The flow is *stable*, and (3.17) can be interpreted as a propagating internal *gravitational wave*. If $\omega = ck$ is inserted in (3.21), then:

$$\omega = N \cos \Psi, \quad \text{with:} \quad N = \sqrt{\frac{g}{T_0} \frac{d\Theta}{dz}},$$

where N is referred to as the *Brunt–Väisälä frequency* and Ψ is the angle between the wave vector (k, m) and the horizontal plane. (Compare this result to Problem 2 of Sect. 2.2.)

2. $d\Theta/dz < 0$

The solution for c is now purely imaginary, indicating that a solution with $c_i > 0$ is always a possibility, in which case the perturbation is unstable. We already encountered this result in Problem 2 of Sect. 3.2, where it was referred to as *convection*.

Physically, case 2 is a situation where a hot fluid exists beneath a colder fluid, which typically represents a situation where a fluid with a lower density exists under a fluid with a higher density. It will be immediately clear that this flow geometry is *unstable*. We refer to this as the *Rayleigh–Taylor instability*. Note that this is fundamentally different from the Kelvin–Helmholtz instability from the previous section.

The next example of a solution to the Taylor–Goldstein equation concerns a base flow with $d\Theta/dz = 0$; thus, no effects occur due to variations in fluid density. We now derive a necessary condition for instability in this flow.

Our point of departure is the assumption that the base flow is defined in the region $z_1 \leqslant z \leqslant z_2$ with homogeneous boundary conditions for the perturbations, that is: $\hat{w} = 0$, or: $d\hat{w}/dz = 0$ for $z = z_1$ and $z = z_2$. We multiply the Taylor–Goldstein equation with the *complex conjugate* $\hat{w}^*$ of $\hat{w}$ (by definition, $\hat{w}\hat{w}^* = |\hat{w}|^2$). We integrate the result for $z_1 \leqslant z \leqslant z_2$, to find:

$$\int_{z_1}^{z_2} \hat{w}^* \left[\frac{d^2\hat{w}}{dz^2} - k^2 \hat{w} - \frac{d^2U/dz^2}{U - c} \hat{w} \right] dz = 0. \tag{3.22}$$

We can rewrite the first term in this expression, with the aid of partial integration and the given boundary conditions, to:

$$\int_{z_1}^{z_2} \hat{w}^* \frac{d^2\hat{w}}{dz^2} dz = \hat{w}^* \frac{d\hat{w}}{dz} \Big|_{z_1}^{z_2} - \int_{z_1}^{z_2} \frac{d\hat{w}^*}{dz} \frac{d\hat{w}}{dz} dz = - \int_{z_1}^{z_2} \left| \frac{d\hat{w}}{dz} \right|^2 dz.$$

Now it follows from (3.22) that:

$$\int_{z_1}^{z_2} \left\{ \left| \frac{d\hat{w}}{dz} \right|^2 + k^2 |\hat{w}|^2 \right\} dz + \int_{z_1}^{z_2} \frac{d^2U/dz^2}{U - c} |\hat{w}|^2 dz = 0. \tag{3.23}$$

Fig. 3.9 Rayleigh's
inflection point criterion
states that it is necessary that
the velocity profile $U(z)$ in a
one-dimensional flow has at
least an inflection point, i.e.
$d^2U/dz^2 = 0$, in order to
become unstable. This is
equivalent to having a
maximum in the amplitude of
the vorticity $\Omega(z)$

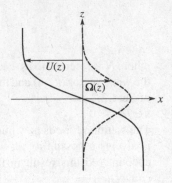

However, the expression in (3.23) is *complex*, since c is also complex. Taking the
imaginary part of this expression, it follows that:

$$c_i \int\limits_{z_1}^{z_2} \frac{d^2U/dz^2}{(U-c_r)^2 + c_i^2}\, |\hat{w}|^2\, dz = 0. \tag{3.24}$$

For *unstable* perturbations, c_i should never become zero. In that case, it follows from
(3.24) that d^2U/dz^2 should change sign somewhere in $[z_1,z_2]$, so that the integral in
(3.24) equals zero. In other words, we just found that $d^2U/dz^2 = 0$ is a *necessary*
condition for instability to occur; in other words, flow instability can only occur when
the velocity profile has an *inflection point*. This is referred to as *Rayleigh's inflection
point criterion*; see Fig. 3.9.

This result is best interpreted in terms of *vorticity*, defined as $\omega_i = \epsilon_{ijk}du_k/dx_j$.
(Please do not confuse here the symbol for vorticity with that of the frequency used
previously in this chapter.) For our one-dimensional flow there is only one non-zero
component of the vorticity: $\Omega(= \omega_y) = -dU/dz$. The condition $d^2U/dz^2 = 0$
means that the amplitude $|\Omega|$ of the vorticity has a *maximum*; see Fig. 3.9.

Some care should be taken here. Rayleigh's inflection point criterion only states a
necessary condition for instability. Later, Fjørtoft found another necessary criterion
for instability, which is a stronger version of Rayleighs theorem.

The inflection point criterion is in fact a generalization of the Kelvin–Helmholtz
instability. This can be clearly seen as the vorticity in the Kelvin–Helmholtz instability is concentrated at the discontinuous velocity interface. We refer to this situation
as a *vortex sheet* or *shear layer*. In other words, the vorticity amplitude in this case
acquires a maximum in the form of a δ-function. In the preceding part of this section it
was demonstrated that a *continuous* vorticity distribution with a maximum amplitude
will also give rise to unstable behavior.

3.3.1 Tollmien–Schlichting Instability

However, we can sense that the presence of an inflection point is not the only criterion for instability. Take for example plane Poiseuille flow between two parallel plates, or boundary layer flow over a flat plate. These are both examples of flows *without* an inflection point, but these flows can certainly become unstable. For an analysis of these types of flows we again assume a base flow with $U = U(z)$ and $d\Theta/dz = 0$. The perturbations applied to this base flow should satisfy the complete set of equations of motions *including* the terms associated with friction. After following an analysis comparable to the one shown above, we arrive at an equation that is comparable to the Taylor–Goldstein equation in (3.20). This equation is called the *Orr–Sommerfield equation*. In dimensionless form, this equation reads:

$$\left(\tilde{U} - \tilde{c}\right)\left\{\frac{d^2\tilde{w}}{d\tilde{z}^2} - \tilde{k}^2\tilde{w}\right\} - \frac{d^2\tilde{U}}{d\tilde{z}^2}\,\tilde{w} = -\frac{i}{\tilde{k}\,Re}\left\{\frac{d^4\tilde{w}}{d\tilde{z}^4} - 2\tilde{k}^2\frac{d^2\tilde{w}}{d\tilde{z}^2} + \tilde{k}^4\tilde{w}\right\}, \quad (3.25)$$

where $\tilde{U}$, $\tilde{c}$, $\tilde{w}$, $\tilde{z}$ and $\tilde{k}$ are dimensionless variables comparable to U, c, $\hat{w}$, z and k, as introduced above. The term Re is a characteristic Reynolds number of the flow. Based on this equation we find that a flow becomes unstable at a certain *critical Reynolds number Re_{cr}*. This is illustrated in Fig. 3.10 for the case of pressure-driven laminar flow between two parallel plates, i.e. plane Poiseuille flow, which has a critical Reynolds number of 5,772 (Drazin and Reid 1981). In the case of a laminar boundary layer flow a similar diagram is found, but with a critical Reynolds number of $1,200$ (based on the boundary layer thickness and the free stream velocity).

This type of instability, related directly to the effect of *viscosity*, is called a *Tollmien–Schlichting instability*. At first glance such a result might be surprising, because intuitively one may expect that friction would dampen the perturbations and thus would have a stabilizing effect. Further on we encounter examples of this more conventional behavior of friction.

Fig. 3.10 Stability diagram of the solutions of the Orr–Sommerfeld equation for a plane channel flow with height $2H$. The shaded region with $c_i > 0$ indicates the domain of unstable solutions. Below the critical Reynolds number $Re_{cr} = 5,772$ (Drazin and Reid 1981) the solutions are stable for all wave numbers k

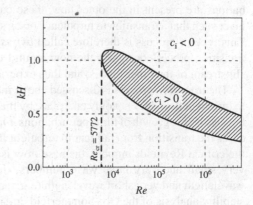

Fig. 3.11 Visualization of a turbulent spot in a boundary layer as seen from above. This is representative of a perturbation that developed from a Tollmien–Schlichting instability. The fluid flow is from *left* to *right*, but the spot develops in the upstream direction (i.e., to the *left*). From: Cantwell et al. (1978)

The instability of Tollmien–Schlichting waves is rather weak. It therefore takes a long time before any infinitesimal perturbation reaches a noticeable amplitude. However, as soon as the perturbations become sufficiently large, a next phase in the instability process begins. Because we are now looking at *finite* perturbations, the effects of the nonlinear terms in the equations of motion start to play a dominant role, and the perturbations grow rather quickly. The flow becomes turbulent, at first limited to some isolated regions that, in the case of a boundary layer, are so-called *turbulent spots*. In Fig. 3.11 such a turbulent spot is shown. The individual spots grow in size while being carried by the flow until they merge into larger structures until the whole boundary layer becomes turbulent.

This *transition scenario*, where a laminar boundary layer flow becomes a turbulent one by means of a Tollmien–Schlichting instability, can *only* occur when very small perturbations are initially present in the flow outside the boundary layer. Such a transition can be observed in the flow over the wing of an airplane. When perturbations are present in the outer flow, these can influence the flow in the boundary layer such that a transition to turbulence occurs by the Tollmien–Schlichting process. This type of process is therefore called *bypass transition*. Another example of this transition process occurs in the flow around turbine blades when a blade is in the slipstream of the other blades, and thus experiences large perturbations.

The transition processes discussed above indicate that the transition from laminar to turbulent flow does not only take place by the growth of infinitesimal perturbations, but also by the growth of finite perturbations. One could even say that in most practical cases the transition from laminar to turbulent flow occurs by this last process. Above the critical Reynolds number, the base flow is linearly stable for perturbations with very small and very large wave numbers, that is perturbations with a very long wavelength and very short wavelength (e.g., surface roughness). Please note that the stability analysis of the Orr–Sommerfeld equation is for *infinitesimal* perturbations; when the perturbations have a large amplitude, then the non-linear terms can still induce unstable behavior and a transition to turbulence. Such a transition is called *sub-critical*. This scenario is important in flows that are insusceptible to *all* infinitesimal perturbations. Well-known examples are plane *Couette flow* between moving parallel plates and axisymmetric Poiseuille flow through a pipe. From experiments we know that these flows do become turbulent. For example, for the Poiseuille flow in a pipe, the empirical critical Reynolds number, defined by the pipe diameter and the average speed, is known to be around 2, 300. However, a theoretical analysis points out that laminar pipe flow is *linearly stable* for *all* infinitesimal perturbations (Drazin and Reid 1981). The emergence of turbulence in these flows can thus only be due to nonlinear growth of finite perturbations. To describe such a transition process we need to set up a nonlinear theory which is, for both flows described above, still an open problem. We return to this at the end of the Sect. 3.4.

3.3.2 Rayleigh Stability Criterion

Another example of stability for a one-dimensional flow is that of a viscous flow between two concentric rotating cylinders (see Fig. 3.12). The inner cylinder has a radius R_i with a rotational speed Ω_i, while the outer cylinder has a radius R_e with a rotational speed Ω_e. The tangential component of the Navier–Stokes equations in cylindrical coordinates reduces to:

$$0 = \nu \frac{\partial}{\partial r} \frac{1}{r} \frac{\partial r v_\theta}{\partial r}.$$

After integration the following velocity profile is obtained:

$$v_\theta(r) = A/r + Br,$$

in which the constants A and B are determined by the boundary conditions at the inner and outer radii, with the result:

$$v_\theta(r) = \frac{1}{R_e^2 - R_i^2} \left[R_i^2 R_e^2 \frac{\Omega_i - \Omega_e}{r} + \left(\Omega_e R_e^2 - \Omega_i R_i^2 \right) r \right].$$

If the radius of the inner cylinder vanishes, i.e. $R_i = 0$, then $v_\theta(r)$ is simply $\Omega_e r$, i.e. a solid body rotation. The pressure gradient in the radial direction is given by the radial component of the Navier–Stokes equation:

$$\frac{v_\theta^2}{r} = \frac{1}{\rho} \frac{\partial p}{\partial r}.$$

To study the stability of the velocity profile between the concentric cylinders the following approach can be followed: Consider a small particle with the same density

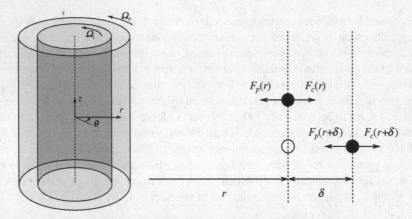

Fig. 3.12 In a rotating flow between two concentric rotating cylinders (*left*) a centrifugal force (F_c) and a pressure force (F_p) act on a particle with the same density as the surrounding fluid (*right*)

as the fluid (i.e., a fluid element) that is rotating with the same velocity as the flow. The forces acting on the particle are the *centrifugal force F_c* and the *pressure force F_p*; see Fig. 3.12. These forces are given by the following expressions:

$$F_c = \frac{\rho v_s^2}{r}\mathcal{V}, \quad \text{and:} \quad F_p = \frac{\rho v_\theta^2}{r}\mathcal{V},$$

in which ρ is the fluid density, v_s the velocity of the particle, v_θ the fluid velocity, r the distance between the center of rotation and the particle location, and $\mathcal{V}$ is the particle volume. It is assumed that initially the particle and fluid velocity are identical, i.e. $v_\theta = v_s$ and thus $F_c = F_p$. Now consider that the particle position is perturbed with an infinitesimal small perturbation δ; see Fig. 3.12. If $F_p(r + \delta) > F_c(r + \delta)$ the particle is pushed back to its original position, and the flow is considered to be *stable*. If $F_p(r + \delta) < F_c(r + \delta)$ the particle is pushed further away from its original position, and the flow is *unstable*.

The forces at the location $r + \delta$ can be written as:

$$F_c(r + \delta) = \frac{\rho[v_s(r + \delta)]^2}{r + \delta}, \quad \text{and:} \quad F_p(r + \delta) = \frac{\rho[v_\theta(r + \delta)]^2}{r + \delta}.$$

The angular momentum of the particle should be conserved, thus: $rv_s(r) = [r + \delta]v_s(r + \delta)$, and $F_c(r + \delta)$ can be written as

$$F_c(r + \delta) = \frac{\rho v_s^2 r^2}{[r + \delta]^3}.$$

The pressure force can be written as:

$$F_p(r + \delta) = \frac{\rho[r + \delta]^2[v_\theta(r + \delta)]^2}{[r + \delta]^3}.$$

The expressions for both forces have now the same denominator, and the flow is thus stable if $v_s^2 r^2 > (r + \delta)^2[v_\theta(r + \delta)]^2$. A Taylor series expansion gives

$$(r + \delta)^2[v_\theta(r + \delta)]^2 \approx r^2[v_\theta(r)]^2 + 2r\delta[v_\theta(r)]^2 + r^2\delta\frac{d}{dr}[v_\theta(r)]^2.$$

The term $r^2[v_\theta(r)]^2$ equals $r^2 v_s$, and the flow is stable if:

$$2\,r\delta[v_\theta(r)]^2 + 2r^2\delta v_\theta(r)\frac{d}{dr}v_\theta(r) > 0, \quad \text{or:} \quad \frac{d}{dr}r v_\theta(r) > 0.$$

The condition for stability is thus: $\frac{d}{dr}(r v_\theta) > 0$. This condition is known as the *Rayleigh stability criterion*. If $v_\theta \propto r^\alpha$ with $\alpha > -1$ the flow is *stable*. For the flow between the two rotating cylinders this implies that for an outer rotating cylinder the flow is *stable*: $v_\theta(r) \approx r$; for an inner rotating cylinder the flow is marginally stable: $v_\theta \approx r^{-1}$, and in practice *unstable*.

Turbulent flows in the presence of a background rotation have a behavior that is quite different than that of turbulent flows in a non-rotating system. Such flows occur in geophysical and astrophysical flows, and in rotating machinery in industrial applications. Rotating turbulent flows are discussed as a special topic in Appendix B.4.

Problems

1. Derive the Taylor–Goldstein equation for perturbations on a one-dimensional inviscid flow.
2. Derive for a flow with a velocity gradient and a temperature gradient, on the basis of the complete set of Taylor–Goldstein equations, that such a flow is stable when the *Richardson number*:

$$Ri \equiv \frac{\frac{g}{T_0}\frac{d\Theta}{dz}}{\left(\frac{dU}{dz}\right)^2} > \frac{1}{4}.$$

Hint: First substitute: $\hat{w} = (U - c)^{1/2}F$, and then derive an equation for F with the first term

$$\frac{d}{dz}\left\{(U - c)\frac{dF}{dz}\right\}.$$

Multiply this equation with F^*, then integrate the result between its boundaries. Then, consider the imaginary part of the integral.

3. Derive the pressure force on a small particle with dimensions dx, dy and dz in a flow with a varying pressure field in one of the coordinate directions.

3.4 Routes to Chaos

In the section above we discussed the conditions for which a laminar flow becomes unstable. This is more or less the first step to a completely turbulent flow. The question that remains: how does complete turbulence emerge? To answer this question we now explore the field of *non-linear dynamical systems* and study how solutions from non-linear deterministic equations, such as the Navier–Stokes equations, make a transition from a stable and ordered regime into an unstable and chaotic regime.

3.4.1 The Logistic Map

One of the simplest non-linear systems is the so-called *logistic map*, defined as:

$$x_{i+1} = \mu x_i (1 - x_i), \quad \text{with: } x_i \in [0, 1], \text{ and: } \mu \in [0, 4]. \tag{3.26}$$

When we substitute:

$$x_i = \frac{u_i + 1}{2}, \quad \text{with: } u_i \in [-1, 1],$$

the logistic map can be re-written in a form that has the equivalent of a non-linear term, a (linear) viscous term, and a forcing term, as in the Navier–Stokes equations:

$$\begin{cases} u_{t+1} - u_t = & -\frac{1}{2}\mu u_t^2 & - & u_t & + (\frac{1}{2}\mu - 1) \\ \dfrac{\partial u_i}{\partial t} = -\left(u_j \dfrac{\partial u_i}{\partial x_j} + \dfrac{1}{\rho_0} \dfrac{\partial p}{\partial x_i} \right) + \nu \dfrac{\partial^2 u_i}{\partial x_j^2} + & f_i \end{cases} \tag{3.27}$$

Therefore, it is sometimes referred to as the *poor man's Navier–Stokes equation* (Frisch 1995), although it has no spatial structure. Nonetheless, it can be used to demonstrate several aspects of non-linear dynamical systems that have their phenomenological counterparts in turbulent flows, such as a transition to turbulence, fractal structure, intermittency, and—most importantly—a finite range of predictability.

Figure 3.13 shows the construction of a *cobweb graph* for the logistic map, where subsequent iterations of (3.26) can be visualized. In Fig. 3.14 cobweb graphs and corresponding transients $\{x_i\}$ are shown for different values of the parameter μ. For $\mu < 1$, the iterations converge to $x = 0$, for *all* initial values $x_0 \in [0, 1]$. In light of the earlier discussion on stability, the solution $x = 0$ is considered to be *stable*. The solution $x = 0$ is called a *stable node* or *fixed point*. This means that once a solution has reached this point, it will remain there forever. When μ becomes larger than 1, there is still a fixed point to which the solution converges, but no longer at $x = 0$, but now at $x = 1 - \mu^{-1}$. This continues for increasing μ, until it reaches a value $\mu = 3$. Then the solution changes its behavior. Now the solution becomes *periodic* and alternates between two values. It is said that the solution has undergone

Fig. 3.13 Construction of a
cobweb graph for the
subsequent mapping
$x_i \longmapsto x_{i+1}$ of the *logistic
map* (3.26); here shown for
$x_0 = 0.3$ and $r = 3.6$

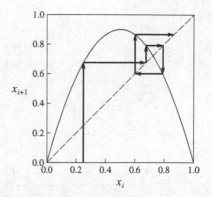

a *bifurcation*. The periodic 'orbit' is stable, and the original fixed point has changed its character, and is now an *unstable* node, or *repeller*. This type of bifurcation is called *super-critical*. The behavior of the solution for increasing values of μ has been depicted in Fig. 3.17.

For $\mu = (1 + \sqrt{6}) = 3.45\ldots$ a second bifurcation occurs. The solution now alternates between *four* values and the period of the 'orbit' has doubled to four. We call this a *period doubling*, which occurs with each bifurcation. When μ further increases, subsequent bifurcations follow up in rapid succession, each time leading to a doubling of the period. At a value of $\mu = \mu_c$, with $\mu_c = 3.57699457\ldots$, the system has undergone an *infinite* number of bifurcations, so that also the period of the orbit has become infinitely long. Hence, we have reached a state where the solution has lost any obvious periodicity.

For $\mu = \mu_c$ it seems at a first glance that the solution is a periodic orbit with period 8. However, closer inspection reveals that each part of the orbit is a 'band' with a finite width, and that each band consists of three narrower bands, which each consist of three yet smaller bands, and so on. This is illustrated in Fig. 3.15. Hence, the orbit is not periodic. Interestingly enough, the same structure is found for different initial values x_0. So, the solution has a degree of stability, although the sequence does not reach a state that is made up of a *finite* number of fixed points. Instead, the state has a *fractal structure* with a dimension of 0.548 (Schuster and Just 2005). In dynamical systems theory such a solution is called a '*strange attractor*'.

The transition from a stationary solution to a periodic one is often encountered as solutions to the Navier–Stokes equations. A well-known example is the formation of a von-Kármán *vortex street* behind a long cylinder. Another example is the formation azimuthal rolls in Taylor–Couette flow at specific flow conditions. Even period doubling is observed when these rolls become modulated in the axial direction.

The subsequent bifurcations suggest a possible transition scenario to reach a 'turbulent' state after an infinite number of bifurcations. The turbulent state is then characterized by a strange attractor. This transition scenario is known as Landau's route to chaos, which is illustrated in Fig. 3.16. Experimentally, Landau's route has never been observed, and in particular the newest discoveries in the area of non-linear

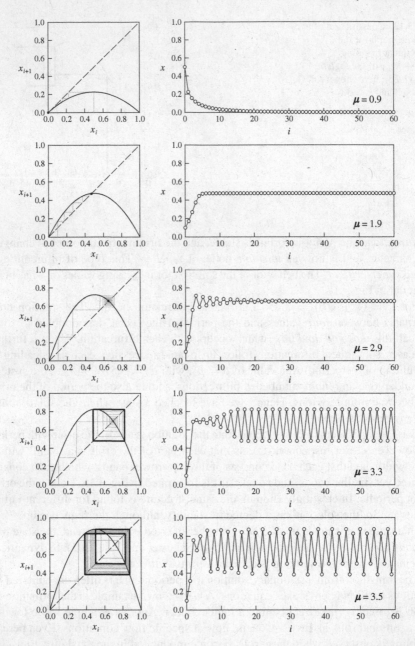

Fig. 3.14 Cobweb graphs and transients of the logistic map (3.26) for increasing values of μ. The transients are drawn as *gray lines* in the cobweb graphs; the fixed points and period orbits are shown as *bold symbols* and *lines*, respectively. When μ increases from 3.3 to 3.5 a period doubling occurs

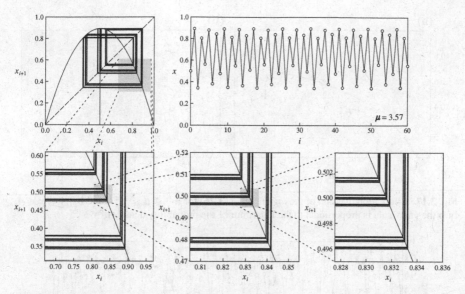

Fig. 3.15 Transient and cobweb graphs of the logistic map for $\mu = 3.5699457...$ There appears to be an orbit with period 8, but it is actually a *strange attractor* with a *fractal* structure

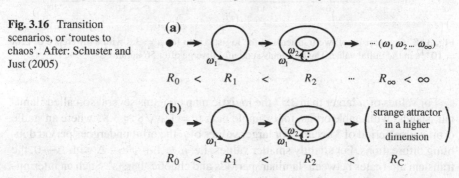

Fig. 3.16 Transition scenarios, or 'routes to chaos'. After: Schuster and Just (2005)

dynamical systems have led to different insights in the emergence of turbulence. We come back to this point later on.

For values $\mu > \mu_c$ the logistic map shows chaotic behavior, apart from some 'islands' with 'laminar' motion. The bifurcation diagram is shown in Fig. 3.17b. However, the orbits are all generated by (3.26), which is purely *deterministic*. This implies that once an initial value x_0 is given, the whole transient for *all* $i > 0$ is *completely* determined. This reminds us of Laplace's quote in the Introduction. However, the predictability of the orbit depends on the precision by which the initial condition has been specified. This is illustrated in Fig. 3.18, where we take two initial conditions that differ by only 10^{-12}. The two solutions are nearly identical for many iterations, but at a certain point the two transients suddenly diverge. The initially very small difference grows exponentially with time; this is further explored in Problem 1.

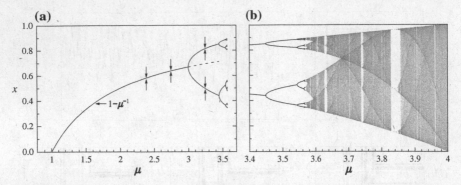

Fig. 3.17 Bifurcation diagram of the *logistic map* (3.26) for **a** $0 \leqslant \mu \lesssim 3.57$, and **b** $3.4 \leqslant \mu \leqslant 4$. In **b** the gray scale is proportional to the logarithm of the probability density of x

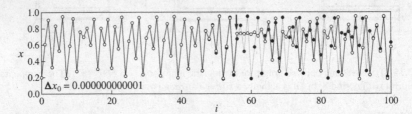

Fig. 3.18 Comparison of two transients of the logistic map for $\mu = 3.8$ with a difference of $\Delta x_0 = 10^{-12}$ in the initial value. The transients suddenly diverge after 56 iterations

For values of μ larger than 3.57 the logistic map contains several so-called 'laminar islands' with stable orbits, for example near $\mu = 1 + \sqrt{8} \approx 3.83$ where an stable orbit with a period of 3 occurs. For larger values of μ the orbit undergoes period doubling bifurcations. For slightly smaller values, i.e. $\mu = 1 + \sqrt{8} - \Delta$ with $\Delta \to 0$, the transient alternates between 'laminar' periods and chaotic 'bursts'. Such an intermittent transient is illustrated in Fig. 3.19. The mean time between these 'bursts' grows proportional to Δ^{-1}. This behavior is explained in Fig. 3.19 by considering the 3-fold logistic map.

This suggests another transition scenario, where the laminar flow is occasionally disrupted by turbulent bursts. As the flow Reynolds number increases, the laminar periods become shorter and shorter, until they eventually vanish and the flow has become fully turbulent. This type of transition is observed for turbulent spots (see Fig. 3.11) in boundary layers, Taylor–Couette flow, and pipe flow. In the case of pipe flow these spots are referred to as 'puffs' (Wygnanski and Champagne 1973) (see also the paragraph 'transition in pipe flow' below).

In conclusion, the logistic map, despite its obvious simplicity, displays very complex behavior that is common to all non-linear dynamical systems. This behavior is phenomenologically similar to that of the Navier–Stokes equations, although it does not provide with a direct explanation, let alone a useable model, for turbulence itself.

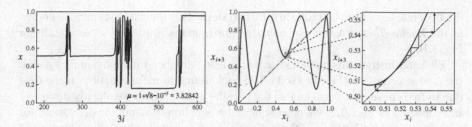

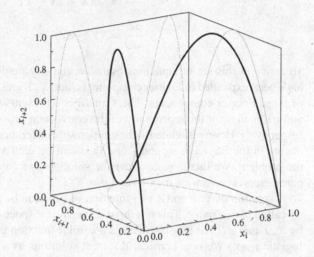

Fig. 3.19 (*left*) The intermittent occurrence of 'laminar' periods and chaotic 'bursts' in a transient of the logistic map with $\mu = 1 + \sqrt{8} - \Delta$. (*middle*) This behavior can be understood when considering the 3-fold map $x_i \longmapsto x_{i+3}$ that nearly intersects the identity mapping; (*right*) the narrow 'passage' requires a large number of iterations, resulting in a 'laminar' period

Fig. 3.20 Plot of the mapping for two consecutive iterations of the logistic map (for $\mu = 4$) shows the '*stretching* and *folding*' that is typical for a non-linear dynamical system

The key feature of non-linear dynamical systems is the so-called '*stretching* and *folding*'. This means that to nearby points always increase their distance, while at the same time the solution is *folded* back onto itself. This is illustrated for the logistic map in Fig. 3.20. We can see multiple examples of such behavior in (turbulent) flows, for example in Fig. 3.5, where the interface between two fluids is first stretched and then folded into vortices.

3.4.2 The Lorenz Equations

It was earlier mentioned that Landau's route to chaos has never been observed experimentally. Instead, it is observed that a flow can undergo only two bifurcations .After the second bifurcation the flow is not periodic anymore, but shows chaotic behavior (characterized by a strange attractor), which we identify with turbulence. In short, the flow reaches a turbulent state in only two steps. This transition scenario is known as the *Ruelle–Takens-Newhouse route to chaos*, and is illustrated schematically in Fig. 3.16.

We can illustrate this transition scenario using another simple dynamic system: the *Lorenz equations* (Lorenz 1963). The background of these equations is the flow due to free convection between two flat parallel plates, which we already discussed in Problem 2 of Sect. 3.2. We present here the equations without the derivation; for a more extensive discussion of these equations we refer to the book by Schuster and Just (2005):

$$\frac{dX}{dt} = -\sigma X + \sigma Y,$$

$$\frac{dY}{dt} = -XZ + rX - Y, \qquad\qquad (3.28)$$

$$\frac{dZ}{dt} = XY - bZ.$$

To arrive at this set of equations both the velocity field and the temperature field have been expanded in Fourier components, and X, Y and Z represent the amplitudes of these Fourier components. The symbol σ represents the Prandtl number, and b and r functions of the aspect ratio of the convection problem and Rayleigh number, respectively. Here, r is related to the temperature difference between the plates (which occurs in the Rayleigh number), and thus can be seen as the main forcing term in the problem. We therefore consider the solution as a function of r, while the other parameters are taken as: $\sigma = 10$ and $b = 8/3$.

A solution of X, Y and Z as a function of time can be presented as a trajectory in so-called *phase space*. This is a three-dimensional space where the axes are defined by X, Y and Z. (The phase space has a similar function as the cobweb graph for the logistic map.) We now consider different solutions as a function of r. This is also illustrated in Fig. 3.21.

For $r < 1$ all trajectories approach the origin in phase space at $X = Y = Z = 0$, which is a fixed point for (3.28). We can interpret this solution as a complete damping of the flow between the two plates. In other words, the heat transport occurs solely through molecular conduction.

For $r = r_1 = 1$ we find the first bifurcation. The fixed point at the origin becomes unstable. This can be shown directly using linear stability analysis; see Fig. 3.22 Two new fixed points emerge at:

$$Z = r - 1, \quad X = Y = \pm\sqrt{b(r-1)}.$$

In physical terms we can interpret the solution represented by these two points as cylindrical flow patterns between the two plates, i.e. so-called *convection rolls*. We find that all trajectories for $r > 1$ approach either of the two fixed points. Please note the two solutions for large t in the two transients for $r = 1.3$ in Fig. 3.21.

We can apply linear stability analysis to the Lorenz equations for $r > 1$. The eigenvalues of the linearized set of equations are shown in Fig. 3.22. For $r = 1$ the

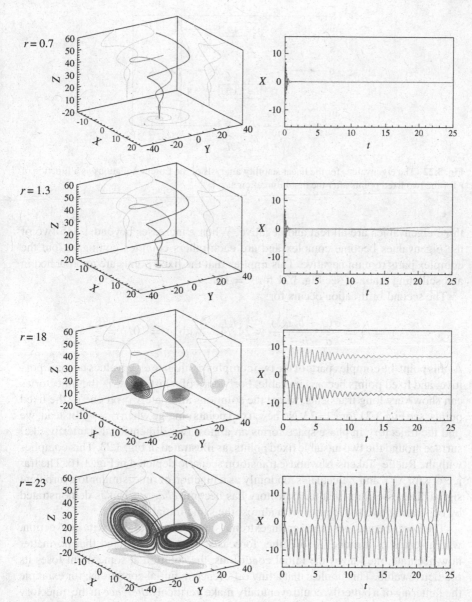

Fig. 3.21 Trajectories in phase space (*left*) and transients of $X(t)$ (*right*) of the Lorenz equations for different values of r (with: $\sigma = 10$, $b = 8/3$) and two symmetric initial conditions: $(10, -10, 60)$ and $(-10, 10, 60)$

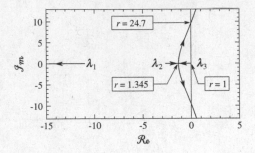

Fig. 3.22 The eigenvalues for the linear stability analysis of the Lorenz equations as a function of r for the two fixed points after the first bifurcation at $r = 1$

three eigenvalues are all real and negative. When r increases beyond 1.345 two of the eigenvalues become complex and are each others complex conjugate, but the complex parts remain negative. This implies that the fixed points are approached in an oscillating manner; see Fig. 3.21 for $r = 18$.

The second bifurcation occurs for:

$$r = r_2 = \frac{\sigma(\sigma + b + 3)}{\sigma - b - 1} \approx 24.74, \quad \text{with:} \, \sigma = 10, b = \frac{8}{3}.$$

At this point the complex parts of the two complex conjugate eigenvalues become positive, and fixed points become unstable. For values of r just below r_2 the trajectories can show very long transients, where the solutions alternately orbit one of the fixed points; see Fig. 3.21 for $r = 23$. No new fixed points emerge when $r > r_2$. Instead, we find the trajectory in phase space forms an almost two-dimensional 'butterfly-like' surface around the two unstable fixed points, as illustrated in Fig. 3.23. This complies with the Ruelle–Takens-Newhouse transition scenario depicted in Fig. 3.16. The trajectory for X, Y and Z fluctuates randomly as a function of time, similar to a turbulent signal. The essence is that the trajectory has become *chaotic*. This is demonstrated by comparing two trajectories with almost the same initial condition.

The Lorenz equations represent an extreme simplification of flow with convection, which serves as a model for weather forecasts. This demonstrated that no matter how accurate one knows the initial conditions, the solution at some point loses its predictive value. This implies that a tiny difference in initial conditions, for example the fluttering of a butterfly, could eventually make a critical *difference* in the trajectory of a hurricane.[1] This is called 'Lorenz's butterfly', which is also the nickname of the strange attractor in Fig. 3.21.

The transition to turbulence for the Lorenz equations follows the scenario of the Ruelle–Takens-Newhouse route to chaos. A very similar transition scenario has been observed in experimental data obtained from Bénard convection; see Fig. 3.24.

---

[1]In popular literature this is often misinterpreted in the sense that hurricanes would be *caused* by the fluttering of butterflies.

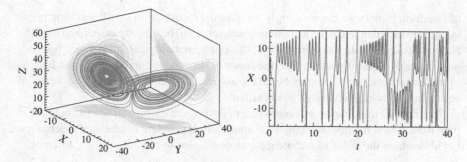

Fig. 3.23 A trajectory given by the Lorenz equations for $r = 28$ (with: $\sigma = 10$, $b = 8/3$). For $r > 24.74$ the Lorenz equations have reached a chaotic state, which is characterized by the *Lorenz attractor*, which here has a fractal dimension of 2.06. The transients are sensitive to the accuracy of the initial conditions, as demonstrated in the right graph, where the difference in initial conditions is 0.000001. The trajectories diverge suddenly at $t \approx 19$

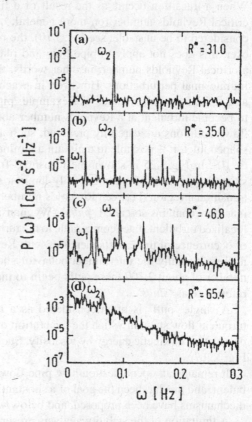

Fig. 3.24 Experimental power spectrum of the velocity measured in Bénard convection by means of laser-Doppler velocimetry (Gollub and Benson 1980). With increasing relative Rayleigh number $Ra^* = Ra/Ra_c$, where Ra_c is the critical Rayleigh number, the following states are observed: **a** periodic motion with a single frequency and its harmonics; **b** quasi-periodic motion with two incommensurate frequencies and their linear combinations; **c** non-periodic chaotic motion with some sharp lines; **d** chaotic motion. After: Schuster and Just (2005)

Yet, there is an important difference between the systems we studied above and the actual Navier–Stokes equations and turbulence. The logistic map and Lorenz equations are examples of so-called low-dimensional systems with only a limited number of degrees of freedom. Turbulent flows have a very large number of degrees of freedom. Hence, we cannot simply transfer the results from non-linear dynamical systems to turbulence. So far, no strange attractor could be directly associated with the Navier–Stokes equations. Nonetheless, the phenomenological similarities between non-linear dynamical systems and (the onset of) turbulence are striking, and help us to understand the very complex behavior associated with a relatively simple set of equations, such as the Navier–Stokes equations.

To complete this chapter we review recent findings in the transition to turbulence in pipe flow, which has not only remained an unsolved problem of classical mechanics, but also where the fields of turbulence and non-linear dynamics strongly interact.

3.4.3   Transition in Pipe Flow

When a transition occurs as the result of a finite-amplitude perturbation below a critical Reynolds number (or, more general, outside the domain where the flow is considered to be unstable; see Fig. 3.10), the transition is called *sub-critical*. However, this does not apply to pipe flow and plane Couette flow, which do not have a critical Reynolds number; in other words, these flows are linearly stable for *all* infinitesimal perturbations. However, in practice these flows do become turbulent at a finite Reynolds number. For example, pipe flow in engineering is considered to become turbulent at a Reynolds number above about 2,300. However, when the flow conditions are carefully prepared, such as a very smooth pipe with a funnel-shaped inlet, it is possible to maintain a laminar flow state up to Reynolds numbers of 10^5. In Fig. 3.25 is shown the measured friction factor (see Sect. 6.3) for transitional pipe flow, between the fully laminar state and fully turbulent state. In the transitional region (between Reynolds numbers of about 2,000 and 2,700) the laminar flow contains so-called '*puffs*' (Wygnanski and Champagne 1973), which are localized turbulent states comparable to the turbulent spot in Fig. 3.11. The frequency of occurrence of these puffs increases with Reynolds number. (This resembles phenomenologically the intermittent behavior shown in Fig. 3.19.) Above a Reynolds number of about 2,700 these puffs begin to increase their length, and they are then referred to as '*slugs*'.

A single 'puff' is often considered as a *minimal flow unit* that can sustain a turbulent flow state, in which the generation of turbulent motion is balanced by the destruction of kinetic energy by viscosity. Such a process is shown schematically in Fig. 3.26.

It remains an open question how pipe flow and plane Couette flow become turbulent, and this has been the goal of a substantial body of research. Several different mechanisms have been proposed, and below we briefly review two of them.

A limitation of the stability analysis presented in the previous sections, is that it only considers the behavior of perturbations in the limit $t \to \infty$. Hence, conventional stability analysis fails to describe the short-term behavior of perturbations. The linearized Navier–Stokes equations typically yield a non-orthogonal set of eigenvectors. This enables large *transient growth* of the perturbation energy, even when all

(a)

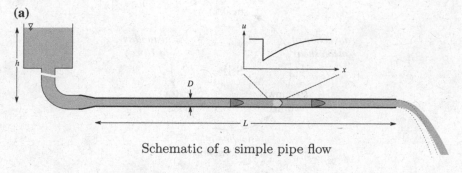

Schematic of a simple pipe flow

(b)

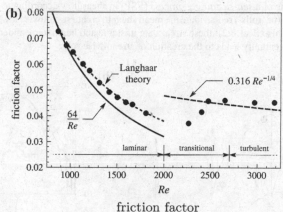

friction factor

(c)

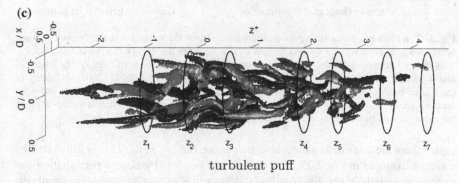

turbulent puff

Fig. 3.25 **a** Schematic of a simple demonstration for transitional pipe flow. The flow is driven by the pressure head in the large container. In the transitional region the jet at the pipe exit intermittently dips, which is the result of the lower velocity in the central region of the pipe in comparison to that of laminar flow. From: Hof et al. (2006) **b** The friction factor as a function of Reynolds number (for a pipe with a 3 mm inner diameter and a length of about 1 m). The *solid line* is the friction coefficient for Hagen–Poiseuille flow; *Langhaar theory* (*dotted line*) describes the friction factor including the development length at the pipe inlet; the *dashed line* represents Blasius' friction law (see Sect. 6.3). The first turbulent puffs appear at the pipe exit for $Re \approx 2{,}260$. **c** Reconstruction of a puff in a pipe flow at $Re = 2{,}000$ from planar stereoscopic PIV data showing streamwise vortices. From: Doorne and Westerweel (2009)

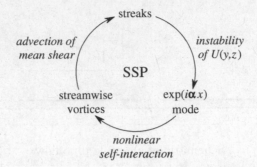

Fig. 3.26 Schematic of a *self-sustaining process* (SSP) in shear flow according to Waleffe (1997): streamwise vortices (or 'rolls') redistribute the mean shear to create streamwise low-speed and high-speed 'streaks' (see also Sect. 8.2); these streaks are unstable and begin to meander with increasing amplitude, which eventually leads to the creation of streamwise rolls

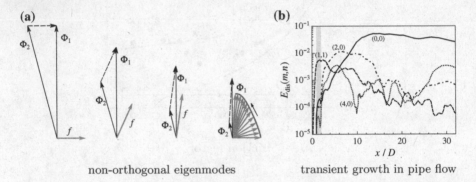

non-orthogonal eigenmodes   transient growth in pipe flow

Fig. 3.27 **a** Example of transient growth. From *left* to *right*: the vector f is the difference between nearly collinear eigenvectors Φ_1 and Φ_2. As time progresses Φ_1 decays at a faster rate than Φ_2. This makes f turn towards Φ_1, while it temporarily grows in length, before decaying. When Φ_1 and Φ_2 would be orthogonal, f would decay monotonically. After: Schmid (2007). **b** Energy of the Fourier modes (m, n) of a perturbation in pipe flow, where m is the azimuthal mode and n the temporal mode. Only the $(1,1)$ mode is imposed in the shaded area. Mode $(0,0)$ is the deviation of the mean velocity profile from the parabolic profile. After: Gavarini et al. (2004)

eigenvalues are confined to the stable half-plane (i.e., $c_i < 0$). This is illustrated by a simple example in Fig. 3.27. The idea should be clear by now: a perturbation can grow substantially in amplitude before it decays; this makes it possible for small perturbations to grow until their amplitude becomes sufficiently large until non-linear effects take over (before the final exponential decay of the perturbation sets in) and cause a transition to turbulence. The initial conditions that maximize the transient growth are referred to as *optimal perturbations*. Typically, such initial conditions contain streamwise vortices, which can be associated with the rolls encountered in Bénard convection and Taylor–Couette flow, and thus have similarities to the Ruelle–Takens-Newhouse route-to-chaos. Such stream wise vortices are also found in puffs in pipe flow (see Fig. 3.25c), and in the turbulent spots that occur in transitional boundary layers (see Fig. 3.11).

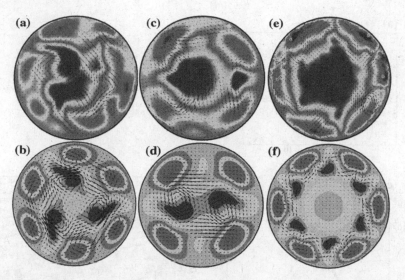

Fig. 3.28 Comparison of traveling wave solutions in pipe flow, obtained in experimental data **a, b, e** and the exact solutions **b, d, f**. The solutions are characterized by their rotational symmetry; shown here are: C3 (**a, b**), C2 (**c, d**), and C6 (**e, f**). From: Hof and van Doorne et al. (2004)

So far, it was assumed that only a single base solution exists for laminar pipe flow, that is the Hagen–Poiseuille flow with a parabolic velocity profile. In 2004 alternative solutions were found of the Navier–Stokes equations that have the form of *traveling waves* (Faisst and Eckhardt, 2003; Wedin and Kerswell, 2004). These alternative solutions are all unstable, so that it is not possible to generate these solutions in an experiment. However, experimental data revealed flow patterns with the same degree of symmetry and a strong qualitative resemblance to the traveling wave solutions; see Fig. 3.28.

The fixed point that represents Poiseuille flow represents a fixed point in phase space, while the family of traveling wave solutions form a *strange repeller*. When a large perturbation is introduced, the dynamic system undergoes a long transient, where the flow solutions wanders among the traveling wave solutions but eventually returns to the fixed point. This is illustrated in Fig. 3.29. First results obtained by both numerical simulations appeared to indicated that these transients would become infinitely long at a certain critical Reynolds number Re_c, which would indicate the transition of the strange repeller into a strange attractor. This behavior was initially confirmed in an actual experiment by considering the lifetime of localized turbulent spots, or 'puffs'. However, later experiments demonstrated that the lifetimes of these transients do not diverge at a finite Reynolds number.

Instead, it appears that puffs split up in multiple puffs above a certain Reynolds number. This happens at an increasing rate with increasing Reynolds number, until the entire pipe volume has become turbulent (Moxey and Barkley 2010; Avila et al. 2011). Figure 3.29d shows the relaxation of a fully-developed turbulent pipe flow

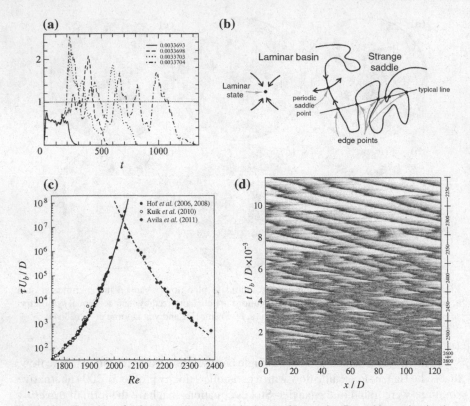

Fig. 3.29 a Transients for finite-amplitude perturbations and increasing Reynolds number in pipe flow; from: Faisst and Eckhardt (2004). **b** Representation of a transient in phase space (graph courtesy of T. Schneider, Univ. of Marburg). **c** The mean life time τ of transients (——) and of puffs before splitting (– – –) as a function of Reynolds number; with data of Hof et al. (2006, 2008); Kuik et al. (2010); Avila et al. (2011). **d** Space-time diagram of the perturbation energy in a pipe flow when fully-developed turbulent pipe flow at $Re = 2,800$ is relaxed to $Re = 2,250$ (time increases from bottom to top of the graph). The flow is from *left* to *right*, but the mean flow speed U_b has been subtracted, so that the puffs (that move slower than the mean flow speed) appear to move backwards (i.e., to the *left*). After: Moxey and Barkley (2010)

at $Re = 2,800$ to a lower Reynolds number of $Re = 2,250$. The gray value along each horizontal line in this graph represents the flow speed at the centerline along the length of the pipe; the darker regions represent turbulent flow regions where the velocity profile is flatter and the centerline flow speed is lower than in Poiseuille flow. As the Reynolds number is lowered in time (time increases from bottom to top), isolated pockets of turbulent flow, or 'puffs' (dark bands) appear that are separated by laminar flow regions (lighter areas). At intermediate Reynolds numbers ($Re \approx$ 2300-2400) one can observe the splitting of individual puffs.

Problems

1. Program the logistic map in MATLAB and determine how many iterations it takes before two nearby transients diverge as a function of the initial difference. The small difference remains undetectably small for a significant time, but grows exponentially with time. Plot the difference in a semi log plot and determine the exponent of the growth rate of the difference between the two transients. This exponent is called the *Lyapunov exponent*.

2. Find the stationary solutions to the Lorenz equations by equating the left-hand side of Eq. (3.28) to zero and solving the resulting algebraic equations. Find the stability of these stationary solutions by applying linear stability analysis.

3. Program the Lorenz equations (3.28) in MATLAB and calculate the behavior as discussed above as a function of r. Determine the growth rate, i.e. Lyapunov exponent, for small differences in the solution of the Lorenz equations in the chaotic regime; see also Problem 1.

4. Continue with Problem 3 and determine the mapping of two subsequent maxima in Z for $r = 28$. Compare this qualitatively with the *logistic map*.

5. Set up a small experiment as shown in Fig. 3.25a. Determine the pressure drop ΔP over the pipe from the difference Δh in elevation between the free surface in the container and the pipe exit. Make a correction for the pressure at the pipe inlet. Determine the (Darcy) friction factor[2], λ from

$$\Delta P = \lambda \frac{L}{D} \frac{1}{2} \rho U_b^2,$$

where L is the pipe length, D the pipe diameter, ρ the fluid density (10^3 kg/m^3 for water), and U_b the mean bulk velocity. Determine an accurate value for D from measurements at low Reynolds number, where $\lambda = 64/Re$. Determine the frequency of puffs that emerge at the pipe exit as a function of Reynolds number.

---

[2]In this book we use the *Darcy friction factor* λ, defined in (6.34); this should not be confused with the alternatively defined *Fanning fraction factor* which is equal to $\lambda/4$.

Chapter 4
The Characteristics of Turbulence

We are now concerned with fully developed turbulence. However, before we derive the equations of motion for turbulent flow in Chap. 5, we summarize in this chapter the most important characteristics and physical properties of turbulent flow.

4.1 The Burgers Equation

Nonlinearity plays an essential role in turbulence. A general solution to the complete set of Navier–Stokes equations is still lacking just because of this nonlinearity. That is why we first focus on a model problem that is on the one hand analytically manageable, and on the other hand contains the essential ingredients of a turbulent flow. This is the *Burgers equation*:

$$\frac{\partial u}{\partial t} + u\frac{\partial u}{\partial x} = \nu\frac{\partial^2 u}{\partial x^2}. \tag{4.1}$$

In this equation we recognize the second term on the left as a nonlinear term, resembling the advection term in the Navier–Stokes equations. The term on the right can be interpreted as friction. Let us consider these two physical processes in the Burgers equation separately. For this, we first focus on the equation:

$$\frac{\partial u}{\partial t} = \nu\frac{\partial^2 u}{\partial x^2}. \tag{4.2}$$

This equation is known as the one-dimensional *diffusion equation*. As an example of a solution to this equation we consider a problem satisfying the following initial and boundary conditions:

$$u = I\delta(x) \quad \text{for: } t = 0, \quad \text{and: } \quad u = 0 \quad \text{for: } x \to \pm\infty \quad \text{and} \quad \forall t. \tag{4.3}$$

© Springer International Publishing Switzerland 2016
F.T.M. Nieuwstadt et al., *Turbulence*, DOI 10.1007/978-3-319-31599-7_4

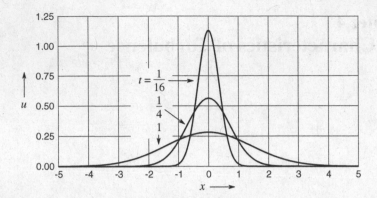

Fig. 4.1 The solution (4.4) of the diffusion equation in (4.2) for the initial and boundary conditions in (4.3) with $I = 1$ and $\nu = 1$. After: Csanady (1973)

where $\delta(x)$ is the Dirac δ-function. The solution reads:

$$u = \frac{I}{2\sqrt{\pi \nu t}} e^{-\frac{x^2}{4\nu t}}, \tag{4.4}$$

which is illustrated in Fig. 4.1 for different values of t. We see that the gradient $\partial u / \partial x$ decreases as t increases. We can thus conclude that friction *suppresses* the gradients in the solution. This has a *stabilizing* effect.

Next we consider the equation

$$\frac{\partial u}{\partial t} + u \frac{\partial u}{\partial x} = 0. \tag{4.5}$$

This is the so-called nonlinear advection equation. The general solution of this equation reads:

$$u = f(x - ut), \tag{4.6}$$

where f is an arbitrary differentiable function. This implies that the value of u for $t = 0$ propagates unaltered through the x, t-surface along a so-called *characteristic*: $x - ut =$ constant. The slope of this characteristic is $dx/dt = u$. In other words, the slope is determined by the value of the solution itself. This leads to an important phenomenon that is illustrated in Fig. 4.2. Here we see that, assuming an initial triangular profile for u, the gradient of the solution, $\partial u / \partial x$, is locally sharpened as a function of time. After a certain time the solution even becomes multi-valued, which would be physically impossible. The conclusion thus reads: the nonlinear term sharpens the gradients in the solution. This has a *destabilizing* effect.

The behavior described by Eq. (4.5), is perhaps best understood using the analogy of shallow water waves, as they develop on the beach. From our own experience we

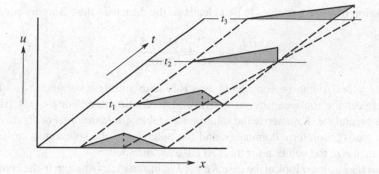

Fig. 4.2 Solution of the nonlinear advection equation in (4.5) for an initial triangular profile for u

Fig. 4.3 Definition of the
macroscopic scales $\mathcal{U}$ and $\mathcal{L}$
in a turbulent flow

know that these waves break. Namely, by these nonlinear effects, the waves become
steeper and steeper, until finally they break and dissipate in the surf.

On the basis of these solutions we can conclude that the solution to the complete
Burgers equation is determined by two opposing processes. The first is due to the
nonlinear term, which is gradient *sharpening*, while the second is due to the diffusive
term, related to friction, which *suppresses* the gradients. It is clear that the ratio of
these processes determines the appearance of the solution.

How can we describe this ratio in a more quantitative manner? To this end, we
write down the Burgers equation in dimensionless form. For this we introduce both
a velocity scale $\mathcal{U}$ and a length scale $\mathcal{L}$, which are illustrated in Fig. 4.3. Next, we
define the following dimensionless variables:

$$\tilde{u} = \frac{u}{\mathcal{U}}, \qquad \tilde{x} = \frac{x}{\mathcal{L}}, \qquad \tilde{t} = t \frac{\mathcal{U}}{\mathcal{L}}. \tag{4.7}$$

Substitution of these variables in (4.1) leads to the dimensionless Burgers equation

$$\frac{\partial \tilde{u}}{\partial \tilde{t}} + \tilde{u}\frac{\partial \tilde{u}}{\partial \tilde{x}} = \frac{1}{Re}\frac{\partial^2 \tilde{u}}{\partial \tilde{x}^2}. \tag{4.8}$$

Here $Re = \mathcal{U}\mathcal{L}/\nu$ can be considered as a Reynolds number, because it yields the ratio between the nonlinear advection term and the friction term. For $Re < 1$, friction dominates, and we characterize the solution as 'stable' (or *laminar*). For $Re \gg 1$, the nonlinear advection term dominates, and the behavior of the solution is 'unstable.' We characterize the solution for this last case as 'turbulent.'

Let us take a closer look at the case $Re \gg 1$ using an exact solution to the complete Burgers equation. This solution reads:

$$u = \frac{\mathcal{U}}{2}\left\{ -\tanh\left(\frac{\mathcal{U}x}{4\nu}\right) + \frac{x}{\mathcal{L}} \right\} \tag{4.9}$$

for $-\mathcal{L} \leqslant x \leqslant \mathcal{L}$, and with:

$$\mathcal{U} = \frac{4U_0}{1 + 2U_0 t/\mathcal{L}}. \tag{4.10}$$

This solution is illustrated in Fig. 4.4 in the case of the characteristic Reynolds number $Re = \mathcal{U}\mathcal{L}/\nu \gg 1$. We find that on the scales of order $\mathcal{O}(\mathcal{L})$ the solution is dominated by the solution of the nonlinear equation (4.5), because in that case we can approximate (4.9) by

$$u = \begin{cases} \frac{\mathcal{U}}{2}\left\{\frac{x}{\mathcal{L}} + 1\right\} \text{ for } x < 0 \\ \frac{\mathcal{U}}{2}\left\{\frac{x}{\mathcal{L}} - 1\right\} \text{ for } x > 0 \end{cases} \tag{4.11}$$

We refer to this as the *macrostructure*. Nonlinear processes dominate the macrostructure and friction is negligible at this scale.

On the other hand, near the origin there is a small region defined by $|x| < 4\delta$, with $\delta = \nu/\mathcal{U}$ where the gradient becomes so large that friction is no longer negligible (see Problem 3a). Here the solution can be approximated by:

$$u = -\frac{\mathcal{U}}{2}\tanh\left(\frac{\eta}{4}\right), \quad \text{with: } \eta = \frac{\mathcal{U}x}{\nu}. \tag{4.12}$$

This is called the *microstructure*, which is dominated by friction.

The Burgers equation now leads to an important insight. In the limit $Re \to \infty$ we can identify both a microstructure and a macrostructure in the solution, each dominated by different physical processes. In particular, we found that the effects of friction shift to the small scales, to the microstructure.

These properties originate from the fact that in (4.8) the Reynolds number multiplies the term with the highest derivative. This means that in the limit $Re \to \infty$ the equation changes its character. In mathematical terms such a limit is called *singular*.

Fig. 4.4 Exact solution (4.9) of the Burgers equation in (4.1) for $U_0 = 1$, $\mathcal{L} = 1$, $\nu = 0.05$, and $t = 1$. The *shaded area* of width $2 \times 4\delta$, with $\delta = \nu/\mathcal{U}$, near the origin indicates the region where friction is dominant

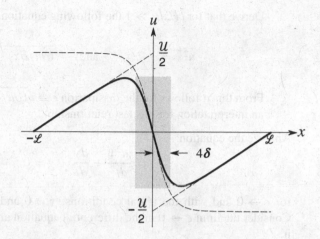

In the following section it is shown that this behavior for the Burgers equation also holds for the complete set of Navier–Stokes equations.

Despite the fact that the Burgers equation provides us with insight into the structure of turbulent flows, as discussed in the next section, we have to admit that the solution to the Burgers equation can not really be considered turbulent; chaotic solutions of this equation simply do not exist. This relates to the fact that the Burgers equation is *integrable*. The meaning of this concept was discussed in Chap. 1.

Problems

1. Derive the solution (4.4) of the diffusion equation (4.2).
2. Rephrase the nonlinear advection equation (4.5) in terms of the gradient $S = \partial u/\partial x$. Derive the following equation for S:

$$\frac{dS}{dt} + S^2 = 0.$$

   Consider the solutions of this equation for $S = S_0$ and $t = 0$. Interpret this solution for $S_0 > 0$ and for $S_0 < 0$.
3. Consider the solution (4.9) of the Burgers equation.

   (a) Show that this equation is an exact solution to the Burgers equation. Consider this solution for $\mathcal{U}\mathcal{L}/\nu \gg 1$. Show that friction is dominant in the region $-4\nu/\mathcal{U} < x < 4\nu/\mathcal{U}$ and that outside this region the solution approximately satisfies the frictionless Burgers equation.

   (b) Define the averaging operator

$$\overline{u}^{\mathcal{L}} = \frac{1}{2\mathcal{L}} \int_{-\mathcal{L}}^{\mathcal{L}} u \, dx.$$

Derive that for $\mathcal{U}\mathcal{L}/\nu \gg 1$ the following equations hold:

$$\overline{u^2}^{\mathcal{L}} = \frac{1}{12}\mathcal{U}^2, \quad \text{and:} \quad \overline{(\partial u/\partial x)^2}^{\mathcal{L}} = \frac{1}{24}\frac{\mathcal{U}^3}{\mathcal{L}\nu}.$$

From this it follows for the dissipation $\epsilon \equiv \nu\overline{(\partial u/\partial x)^2} = \frac{1}{24}\mathcal{U}^3/\mathcal{L}$. Provide an interpretation for this last relationship.

4. Analyze the equation:

$$\epsilon\frac{d^2y}{dt^2} + \frac{dy}{dt} + y = 0,$$

for $\epsilon \to 0$ and with the initial conditions: $y = 0$ and $dy/dt = e^{-1}$ for $t = 0$. Consider the limit $\epsilon \to 0$ in the differential equation and in the exact solution to it.

4.2   Phenomenology

We now discuss the characteristics and properties of turbulent flows that are solutions to the full three-dimensional Navier–Stokes equations. It is best to base a phenomenological description on our own observations. For this we take the famous drawing by Leonardo da Vinci, shown in Fig. 4.5. This represents a turbulent flow of water issuing into a reservoir. What can we see in this drawing?

The flow appears to be *chaotic*, but certainly not *random* if we define random as 'completely determined by chance.' Indeed, we see in Fig. 4.5 that turbulence exists in vortex-like structures, often called *eddies*. Consequently, the velocity measured in two nearby points is correlated as a function of the distance between these positions,

Fig. 4.5 Drawing of water issuing into a reservoir by Leonardo da Vinci (1507)

but also as a function of time. We would never find such a correlation (which will be elaborated more in Chap. 9) in a process that is purely random. Thus, our first observation reads:

Turbulence consists of chaotic vortex-like structures with varying dimensions.

It has to be emphasized however, that not every chaotic flow is turbulent (think for example of waves at the surface of the ocean or acoustic noise).

We refer to the turbulence structure with the largest dimensions as the *macrostructure*. The macrostructure is associated with a length scale $\mathcal{L}$ and with a velocity scale $\mathcal{U}$. These scales are related directly to the geometry of the flow. We later see that we can characterize turbulence at first by $\mathcal{U}$ and $\mathcal{L}$. The reader should not be surprised to learn that the smallest dimensions in Fig. 4.5 are referred to as the *microstructure*, which is further elaborated later in this book.

We now line up the properties of turbulent flow, keeping in mind the solutions of the Burgers equation, as discussed in the previous section.

First of all, the Reynolds number for a turbulent flow is very large, i.e. $Re = \mathcal{U}\mathcal{L}/\nu \gg 1$. This means that nonlinear processes dominate the macrostructure, while viscous effects are negligible at this scale. The macrostructure is thus described essentially by the limit $Re \to \infty$. Consequently, the macrostructure is independent of the Reynolds number. The only relevant parameter ν, which is determined by the fluid, is contained in the Reynolds number, and thus we find that the large-scale structure is independent of ν. We call this *Reynolds similarity*, and this is concisely expressed as:

Turbulence is a property of the *flow*, not of the fluid.

Secondly, we know from experience that turbulence is strongly *dissipative* and *diffusive*. By 'dissipative' we mean that a turbulent flow quickly loses its kinetic energy, and thus quickly decays (provided we do not continually add energy to the flow). We call the kinetic energy per unit mass e, and this energy e scales as $e \sim \mathcal{U}^2$, which indicates that the macrostructure contains most of the kinetic energy. For e the following equation holds, which is derived in Chap. 7,

$$\frac{de}{dt} = -\epsilon, \tag{4.13}$$

where ϵ represents the *dissipation rate* of kinetic energy. One of the most important results of turbulence theory now reads:

$$\epsilon \propto \frac{\mathcal{U}^3}{\mathcal{L}}. \tag{4.14}$$

In other words, the dissipation rate scales with the macrostructure. This is referred to as the Kolmogorov relation, which we can interpret as follows: a turbulent eddy with energy $\sim \mathcal{U}^2$ either loses its energy in one time scale $\mathcal{T} \sim \mathcal{L}/\mathcal{U}$, or the eddy breaks up in a period $\sim \mathcal{T}$ by instability into smaller eddies.

We now focus on the diffusivity of turbulence. From practice we know that turbulent flows are very effective in mixing a fluid. Take as an example stirring your coffee with milk, a mixing process used on a daily basis. Now effective mixing is related to the macrostructure. The larger eddies stir the flow at a scale $\mathcal{L}$, and we expect that the flow variables are transported and mixed at this same scale. Now suppose that diffusion in turbulent flow can be described using a *diffusion equation*, that is

$$\frac{\partial \chi}{\partial t} = K \frac{\partial^2 \chi}{\partial x^2}, \tag{4.15}$$

for the concentration χ. The solution to this equation is discussed in Sect. 4.1, and from this it follows that K plays the role of a *diffusion coefficient*. In this case we refer to K as a *turbulent* diffusion coefficient, or *exchange coefficient*, and in the light of our discussion above, we can scale it as follows:

$$K \sim \mathcal{U}\mathcal{L}. \tag{4.16}$$

It is now easy to see that K is much larger than the *molecular* diffusion coefficient κ; in the case that the Schmidt number[1] is $Sc \equiv \nu/\mathbb{D} \approx 1$, it follows that $K/\mathbb{D} \approx Re \gg 1$.

Finally, the most important property of turbulence in terms of its dynamics is *vorticity*. Thus, the most compact definition of turbulence should read:

Turbulence is 'chaotic vorticity.'

The importance of vorticity could have been expected given our discussion of the Kelvin–Helmholtz instability in terms of vorticity. However, a complete discussion on the role of vorticity in turbulence is postponed till Chap. 8.

So far we have only briefly discussed some properties of turbulence related to its macrostructure. Now we focus to the microstructure. We have seen that the macrostructure loses its energy through instability processes within one characteristic time scale. Obviously, the kinetic energy can only be dissipated, i.e. transformed into heat, by viscosity. This viscous dissipation takes place at the microscale, because the gradient of the velocity can become so large at these small scales that friction can no longer be neglected. Please take recourse to our discussion on the role of the microstructure in the solution of the Burgers equation.

How can we characterize the microstructure in terms of scaling parameters? First, an essential parameter should be the viscosity ν. Second, we expect that $\mathcal{U}$ and $\mathcal{L}$, which describe the macroscale, are not of direct importance to the microstructure, because the information of the macrostructure would, as it were, be lost in the instability processes where larger eddies break up into smaller eddies. The only parameter of physical importance to the microstructure is the amount of energy per unit time that is dissipated. This is characterized by ϵ. Thus, the scaling parameters for the

---

[1] In case we consider the diffusion of heat with a molecular diffusion coefficient α, χ refers to the temperature, and the ratio $Pr = \nu/\alpha$ is the so-called *Prandtl number*.

microstructure are ν and ϵ. With this the following scales for the length, time, and velocity can be defined as:

$$\eta = \left(\frac{\nu^3}{\epsilon}\right)^{\frac{1}{4}}, \qquad \tau = \left(\frac{\nu}{\epsilon}\right)^{\frac{1}{2}}, \qquad v = (\nu\epsilon)^{\frac{1}{4}}, \tag{4.17}$$

respectively. These are known as the *Kolmogorov* scales. On the basis of the equations mentioned above, it follows that the Reynolds number for the *microstructure* is $Re = v\eta/\nu = 1$. Again, this reflects how the microstructure is dominated by friction.

Combining the expressions (4.14) and (4.17) gives the following scaling for the macroscopic length scale $\mathcal{L}$ relative to the Kolmogorov length scale η that represents the microscale:

$$\frac{\mathcal{L}}{\eta} \sim Re^{3/4}. \tag{4.18}$$

This scaling implies that $\mathcal{L} \gg \eta$ for large enough Re. This is an important result, as it represents the dynamic decoupling of the macrostructure and microstructure, provided that the Reynolds number is sufficiently large. This is further explained in Chaps. 8 and 9.

The complete picture now becomes clear. We identified a macrostructure, which is fed by energy drawn from the average flow through instability processes. These energetic large eddies are in their turn unstable and break up into smaller eddies. This process is known as the *energy cascade*. The breakup of larger eddies into smaller ones continues until the microstructure has been reached. There, the kinetic energy is dissipated into heat through viscous friction. The energy cascade is summarized by the following famous verse by L.F. Richardson (1922):

> Big whirls have little whirls,
> That feed on their velocity;
> And little whirls have smaller whirls
> And so on to viscosity

Table 4.1 summarizes the properties of the macrostructure and the microstructure. Finally, Fig. 4.6 illustrates how the macrostructure and the microstructure form the ingredients of turbulence.

Problems

1. Give an estimate of the dissipation of the turbulence in a cumulus cloud, both per unit mass and for the total cloud. Base this estimation on the velocity scale and length scale of a typical cumulus cloud. What are the magnitudes of the Kolmogorov scales?
2. Figure 4.7 shows the turbulent flow behind a grid. The turbulence is advected downstream at a constant mean velocity U_0. Consider a square box that moves along with the flow at a velocity U_0 with a characteristic size $\mathcal{L}$. There is no energy source, so that the turbulence eventually decays.

Table 4.1 Summary of the properties of the macrostructure and the microstructure in turbulence

Macrostructure		Microstructure
– Produced by the average flow (active)	Energy $\longrightarrow$ Cascade ϵ	– Fed by the cascade process (passive)
– Frictionless		– Dominated by friction
– Effective transport of: mass, momentum and energy		– Molecular transport processes
– Depends on flow geometry		– Universal
– Anisotropic		– Isotropic
– Scales: $\mathcal{U}$, $\mathcal{L}$		– Scales: v, η

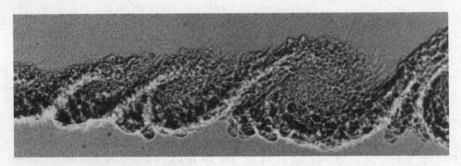

Fig. 4.6 Visualization of the density variations in a turbulent mixing layer, taken by Brown and Roshko (1974), showing the macrostructure and microstructure in a turbulent flow. From: Van Dyke (1982)

(a) For the turbulent length scale we can take $\mathcal{L}$. Explain why. Derive an expression for the behavior of the kinetic energy as a function of time on the basis of the equation $de/dt = -\epsilon$.

(b) Whenever the Reynolds number $\mathcal{U}\mathcal{L}/\nu$ (where $\mathcal{U} = \sqrt{\frac{2}{3}e}$ is a characteristic velocity) becomes less than 10, the estimate of the dissipation ϵ needs to be adjusted to $\epsilon = c\nu\mathcal{U}^2/\mathcal{L}^2$. Provide arguments for this. Calculate the constant c given that the dissipation rate is continuous for $\mathcal{U}\mathcal{L}/\nu = 10$. How does the turbulent energy decrease in this so-called final period of decay? Calculate a numerical example with $\mathcal{L} = 1\,\text{m}$, $\nu = 1.5 \times 10^{-5}\,\text{m}^2/\text{s}$ and $\mathcal{U} = 1\,\text{m/s}$.

(c) Until now we have neglected the effects of the walls of the box. Can this assumption be supported?

3. We assemble a hot-wire anemometer (see Sect. 4.3) on the nose of an airplane to measure the fluctuations of the velocity while the airplane flies through a turbulent atmosphere at a speed of 50 m/s. The length scale and velocity scale of the large vortices are 100 m and 0.5 m/s, respectively. What is the highest frequency of the velocity fluctuations that we measure with the hot-wire? What is the maximum

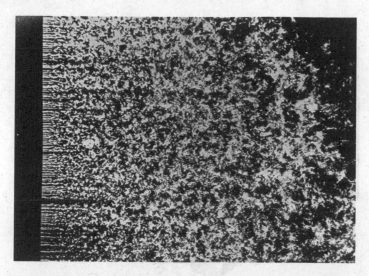

Fig. 4.7 Turbulent flow behind a grid. The turbulence gradually decays as it is convected downstream. The smallest scales decay first, while energy is transferred from the larger scales to the smaller scales. Image courtesy of H. Nagib, Illinois Institute of Technology

length of this hot-wire if we want to measure the complete turbulence spectrum? Estimate the minimum signal-to-noise level of the electronic measurement circuit that is needed for these measurements?

4. In a turbulent flow of a fluid with viscosity ν the largest eddies have the size $\mathcal{L}$ and characteristic velocity $\mathcal{U}$. These eddies break up into multiple smaller eddies with a size $\alpha\mathcal{L}$ (with $\alpha < 1$). The number of these smaller eddies is $n < \alpha^{-3}$. This process repeats itself for the smaller eddies, forming a cascade process.

(a) Calculate the energy of the eddies at the subsequent scales in this cascade process.

(b) How many steps are required for this cascade process until the microstructure is reached? Calculate the Kolmogorov scales and the dissipation rate. How long does it take until an eddy has run through the complete cascade process?

(c) At the end of this process, the distribution of energy over space is as illustrated in Fig. 4.8. Such an object is called a 'fractal.' Determine the spatial dimension of this fractal. Why would the fractal model for turbulence be unrealistic?

5. Compare the visualizations of a turbulent mixing layer (see Sect. 6.6) in Figs. 4.6 and 6.16. Which of the two flows has the highest Reynolds number?

Fig. 4.8 A fractal as a model of the spatial energy distribution in turbulence. From: Mandelbrot (1977)

4.3 Experimental and Numerical Methods

In order to characterize turbulent flows one can either use an experimental or numerical method. Experimental methods generally provide measurements of one or more components of the velocity in a point, a plane, or even a volume. Mostly, experimental data provide statistical information on the turbulent flow (see Chap. 5) or provide insight in the instantaneous flow to reveal vortical flow structures (see Chap. 8). However, most experimental methods provide only an incomplete picture of the flow, either in terms of a reduced number of velocity components, a reduced spatial representation, a limited spatial resolution, or limited flow information (e.g., velocity data, without data on the pressure or temperature of the flow). Numerical methods aim to solve the Navier–Stokes equations, either the full set of equations, or a reduced set of equations including modeling part of the turbulent flow. In this chapter we describe *direct numerical simulation*, or DNS, which is a full numerical simulation of a turbulent flow that does not rely on any modeling of the flow. However, this approach is computationally intensive, and it requires substantial computing facilities. Hence, for many practical problems one has to rely on turbulence models, which is described in later chapters.

Flow Visualization

Many qualitative aspects of flows, also turbulent flows, can be observed by means of *flow visualization*. Mostly, fluid motion is not directly visible, as common fluids, such as air and water, are optically transparent. In order to visualize the motions in the fluid, it is necessary to add a dye, smoke, or small particles to the flow that reveal the fluid motion; it is important to introduce these materials at specific locations to observe specific features of the flow. The images in Figs. 3.7 and 3.11 were obtained by adding aluminium flakes to the fluid (silicone oil and water, respectively), whereas the images shown in Figs. 3.5, 4.3, 6.17, and 6.21a where obtained by means of adding a dye to the (turbulent) fluid. Similar results can be obtained in gas flows (i.e., air) by means of adding smoke (Figs. 6.21b and 8.7), or a fog of tiny droplets of water (Fig. 2.1) or oil (Fig. 6.7). Tracers, in the form of tiny hydrogen bubbles (with a typical diameter of a few tens of micrometers), can be introduced into a water flow by means of *electrolysis* from a thin wire at a specific location in the flow; an example is shown in Fig. 8.5.

Other approaches are so-called *shadowgraphy* and *Schlieren imaging* that visualize small variations in refractive index (caused by differences in density or solution concentration), which causes refraction of light that passes through the fluid. Examples of shadowgraphs and Schlieren imaging are shown in Figs. 4.6, 6.16, and 6.23.

For technical details of these different approaches we refer to the book by Merzkirch (1987). A collection of flow visualization images can be found in the book by Van Dyke (1982).

Single-Point Measurements

Most conventional measurement methods in fluid mechanics, and in particular for turbulent flows, determine the flow velocity (either directly or indirectly) in a single point as a function of time. These methods are: the *Pitot tube*, the *hot-wire anemometer* (HWA), and the *laser-Doppler velocimeter* (LDV). Here we limit ourselves to a very concise description; for details on these methods we refer to any review paper, text book or reference book on experimental methods in fluid mechanics (e.g., Comte-Bellot 1976; Goldstein 1996; Tropea et al. 2007).

A Pitot tube measures the mean difference between the static and dynamic pressure $\overline{\Delta p}$, which is proportional to:

$$\overline{\Delta p} = \frac{1}{2}\rho \left(\overline{u_1}^2 + \overline{u_1'^2} \right),$$

in case the probe is aligned with the x_1-axis. For most turbulent flows, we have: $\overline{u_1}^2 \ll \overline{u_1'^2}$, so that the measurement yields essentially the mean velocity. The *superpipe* data shown in Fig. 6.3 were determined by means of accurate Pitot-tube measurements. A major limitation is that only the mean flow velocity can be determined.

However, a strongly fluctuating velocity is characteristic for turbulence. A hot-wire anemometer is ideally suited for the measurement of these rapid fluctuations. The measurement principle is based on the fact that the convective heat transfer of an electrically heated thin wire depends on the flow speed of a gas (i.e., air) that passes over the wire. The cooling of the wire reduces the electrical resistance. The most common implementation uses an electrical circuit, which consists of a Wheatstone bridge and differential amplifier, that maintains a constant wire temperature.[2] Then,

---

[2]Hence, this implementation is often referred to as *constant temperature anemometry* (CTA).

the voltage E is related to the flow velocity U by:

$$E^2 \cong (T_w - T_0)\left[A + BU^n\right], \quad \text{or:} \quad U \cong C_0 + C_1 E + C_2 E^2 + C_3 E^3 + C_4 E^4 + C_5 E^5,$$

with: $n \cong 0.5$ (typical value), and where T_w and T_0 are the wire temperature and gas temperature respectively, and A, B, and $\{C_0, C_1, \ldots, C_5\}$ are calibration constants. The wire is typically a millimeter or so in length and has a thickness around $5\,\mu\text{m}$, and is made of tungsten (for its high strength). The wires are most sensitive to the flow normal to the wire, and therefore require careful calibration. This arrangement yields a very high frequency response of around $100\,\text{kHz}$, which can resolve even the smallest variations in velocity, proportional to frequencies of U_0/η (where U_0 is the advection velocity by which the turbulent flow passes the wire; see Sect. 9.4). Hence, this makes HWA ideally suited for the measurement of the spectral density of the velocity fluctuations (see Chap. 9). So-called X-wire probes (Comte-Bellot 1976) can measure the correlation between two fluctuating velocity components, or *Reynolds stress* introduced in Sect. 5.3. Also, two-wire probes exist that measure the correlation of the same velocity component but separated in space; this is further elaborated in Chap. 5. Complicated multi-wire probes were developed that can determine instantaneous spatial gradients of the velocity, which reveals one or more components of the deformation tensor and vorticity vector (Wallace and Foss 1995).

Many of the experimental results shown in this book were obtained by means of HWA; for example, most of the data in Figs. 6.19, 9.4, 9.9, and 9.14.

Laser-Doppler velocimetry (or, sometimes called laser-Doppler *anemometry*) uses the physical phenomenon that the frequency of light reflected by a moving object is shifted in proportion to the velocity of the object toward or away from the observer; this is known as the *Doppler effect*. In most approaches in laser-Doppler velocimetry, or LDV, a *heterodyne* approach is used where two laser beams cross each other at an angle θ in a small measurement volume; see Fig. 4.9. The two beams can be thought of to create an interference pattern with a fringe spacing d_f given by:

$$d_f = \frac{\lambda_0}{2\sin(\theta/2)}, \tag{4.19}$$

where λ_0 is the light wavelength. When a small tracer particle passes through the fringes in the small measurement volume, the light that is scattered by the particle fluctuates with a frequency $f = u_\perp/d_f$, where $u_\perp$ is the component of the particle velocity normal to the planes of the fringes. The modulated scattered light from the small particles passing through the measurement volume is detected by a sensitive photomultiplier. Typical laser beam angles are 10–$12°$, giving a typical fringe spacing of a few micrometers and a light modulation frequency of $500\,\text{kHz}$ for a particle with a velocity of $1\,\text{m/s}$. Typical dimensions of the small tracer particles, which are either naturally present in the flow or have been added for this purpose, are 1–$10\,\mu\text{m}$. The small size ensures that the particles accurately follow the fluid motion, even when there is a considerable difference in density between the particle material and fluid.

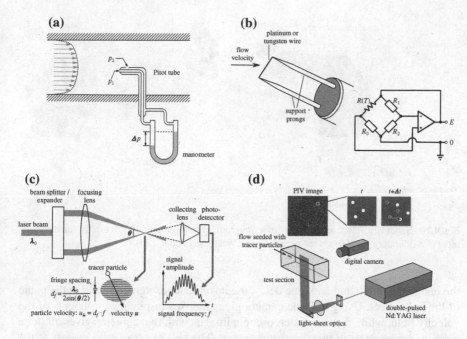

Fig. 4.9 Schematics of common measurement methods for turbulent flows. Images **a–d** are adapted from those at: www.engineeringtoolbox.com, Chen et al. (2003), www.newton.ex.ac.uk, www.thermopedia.com, and Westerweel et al. (2009). **a** Pitot tube. **b** Hot-wire anemometry (HWA). **c** Laser-Doppler velocimetry (LDV). **d** Particle image velocimetry (PIV)

The typical dimensions of a measurement volume is $0.1 \times 0.1 \times 1\,\text{mm}^3$; hence, the velocity is effectively measured in a *point*.

Using two laser beam pairs, with the plane defined by one pair normal to that of the other pair, it is possible to measure simultaneously two velocity components.[3] This enables the direct measurement of the *Reynolds stress* (see Sect. 5.3) in a much simpler manner than with an X-wire HWA probe. Figures 5.2, 6.2, 6.4, and 6.5 are examples of experimental data obtained with LDV.

Particle Image Velocimetry

In *particle image velocimetry*, or PIV, the flow is seeded with small tracer particles that accurately follow the fluid motion. Typically, a planar cross section of the flow is illuminated with a thin *light sheet* that is generated from a dual pulsed-laser system. The light scattered from the particles in the light sheet is recorded by a camera in two subsequent image frames (one for each laser light pulse from the dual laser system). The images are subdivided in small interrogation sub-images, and the displacement of the particle images between the two laser pulses, separated by a time delay Δt, is determined in each sub-images. This gives, after dividing the displacements by

---

[3]Dedicated LDV systems that use three laser beam pairs can measure all three velocity components.

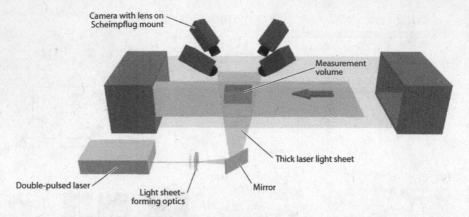

Camera with lens on
Scheimpflug mount

Measurement
volume

Thick laser light sheet

Double-pulsed laser

Light sheet–
forming optics

Mirror

Fig. 4.10 Schematic of the arrangement of a four-camera tomographic PIV setup for the measurement of a boundary layer in a wind tunnel. From: Westerweel et al. (2013)

the image magnification and time delay, the instantaneous velocity field in the plane of the light sheet. Using a single camera gives the two in-plane components of the velocity field, while the use of stereoscopic images with two cameras gives all three velocity components in the observation plane. The two in-plane components of the velocity make it possible to determine the *vorticity* component that is normal to the measurement plane. This is key to understanding the dynamics of turbulent flows (see Chap. 8) and the identification of coherent flow structures, such as hairpin vortices (see Fig. 8.6). Examples of velocity fields measured with PIV are shown in Figs. 5.1 and 8.8. Experimental data obtained with PIV can be found in Figs. 6.19, 6.20, 6.22, and 9.11 (Fig. 4.10).

In *tomographic* PIV a flow *volume* is illuminated, rather than a plane, and four or more cameras are used to record the particle images. From these data the original particle positions within the measurement volume are computed by means of a tomographic reconstruction. Although time consuming, this provides the full volumetric measurement of all three components of the velocity field in the measurement volume. This makes it possible to extract *all* nine components of the deformation rate tensor, and enables the experimental evaluation of the *vortex stretching term* in Eq. (8.4).

A typical spatial resolution for PIV is $1 \times 1 \times 1\,\text{mm}^3$ for each velocity measurement, with an accuracy of about 1 % for the velocity.[4] Large-format electronic image sensors (containing 16 million *pixels*) can yield around 250,000 velocity measurements per image. The use of high-speed digital cameras make it possible to record time-resolved sequences of turbulent flows. For further reading on PIV we recommend the book by Adrian and Westerweel (2011).

---

[4]The relative error for the velocity is based on the displacement measurement error relative to the (nominal) maximum measurable displacement in each sub-image domain.

Direct Numerical Simulation

In a *direct numerical simulation* (DNS) the equations of motion (2.20–2.22) are integrated numerically to solve the velocity, pressure, and temperature fields for given initial and boundary conditions. Here we provide a brief description of the DNS method; for further reading we suggest the review paper by Moin and Mahesh (1998).

The objective of a DNS is to solve the complete set of equations of motion without relying on any modeling of the turbulent flow. This implies that both the macrostructure and the microstructure are resolved by the numerical grid on which the flow is computed and by the temporal integration of the discretized equations. Suppose that the numerical grid has a spatial resolution Δ, given by:

$$\Delta \equiv (\Delta x \, \Delta y \, \Delta z)^{1/3}, \tag{4.20}$$

where Δx, Δy, and Δz are the dimensions of a grid cell in the x, y, and z directions, respectively. In order to resolve both the macrostructure and the microstructure, it is required that the simulation domain is large enough to contain the large-scale motions of a size $\mathcal{L}$, while at the same time the spatial resolution Δ should be small enough to resolve the Kolmogorov scale η. Given the scaling in (4.18), this implies that the total number of grid cells $\mathcal{N}$ in the simulation should be:

$$\mathcal{N} \approx \left(\frac{\mathcal{L}}{\eta}\right)^3 = \mathcal{O}\left(Re^{9/4}\right). \tag{4.21}$$

Furthermore, the integration time step Δt should be small enough to resolve the Kolmogorov timescale τ, while the total integration should be continued sufficiently long to cover a certain number of integral timescales $\mathcal{T}$. Given that $\mathcal{T} = \mathcal{L}/\mathcal{U}$, it can be shown that:

$$\frac{\mathcal{T}}{\tau} \sim Re^{1/2}, \tag{4.22}$$

so that the total *computational effort*, that is the total number of computations required, scales according to:

$$\mathcal{N} \times \mathcal{M} = \mathcal{O}\left(Re^{9/4} \cdot Re^{1/2}\right) = \mathcal{O}\left(Re^{11/4}\right), \tag{4.23}$$

where $\mathcal{M}$ is the total number of time integration steps. Hence, the computational effort for a DNS grows almost with the cube of the Reynolds number. This is quite unfavorable, and therefore DNS has been primarily limited to rather low to moderate Reynolds numbers and simplified flow geometries, such as channels and pipes. Such simplified and generic flow geometries allow the use of *periodic* boundary conditions (BC) in one or more principal directions; examples are: isotropic turbulence (3 periodic BC), pipe and channel flows (2 periodic BC), and boundary-layer flow (1 periodic BC). These periodic boundary conditions represent quasi-infinite domains. Also, this makes it possible to represent the equations and solution as *spectral modes*, which not only reduces the computational effort, but also improves the accuracy and spatial resolution of the simulation.

Spatial and Temporal Discretization

The spatial discretization should be as accurate as possible for the whole range of scales and not add artificial diffusion to the flow. A first-order discretization of the non-linear term, here for simplicity represented by f, gives the following:

$$\frac{\partial f}{\partial x} = \frac{f(x) - f(x - \Delta x)}{\Delta x} + \frac{1}{2}\Delta x \frac{\partial^2 f}{\partial x^2}. \tag{4.24}$$

In this equation Δx is the grid spacing. The truncation error is proportional to the second derivative of f, and the finite grid spacing Δx can be interpreted as an additional viscosity. This additional viscosity leads to a non-physical damping of the turbulence. Higher-order upwind methods have in general similar problems. Instead, central-difference numerical schemes using $x + \Delta x$ and $x - \Delta x$ have a truncation error that is proportional to the third derivative of f and do not introduce an additional viscosity. The first derivative of f is then written as:

$$\frac{\partial f}{\partial x} = \frac{f(x + \Delta x) - f(x - \Delta x)}{2\Delta x} + \mathcal{O}\left(\Delta x^2\right). \tag{4.25}$$

If periodicity of f is assumed, then f can be written as a sum of *spectral modes*, that is

$$f(x) = \sum \hat{f}_k \exp(ikx),$$

where: $i = \sqrt{-1}$, and k is the wave number. This gives for mode $\hat{f}_k$:

$$ik\hat{f}_k \exp(ikx) = \hat{f}_k \frac{\exp[ik(x + \Delta x)] - \exp[ik(x - \Delta x)]}{2\Delta x} + \mathcal{O}(\Delta x^2). \tag{4.26}$$

Thus,

$$\text{error} = \mathcal{O}(\Delta x^2) = ik\hat{f}_k \exp(ikx)\left(1 - \frac{\sin[k\Delta x]}{k\Delta x}\right). \tag{4.27}$$

The error depends thus on the wave number k. For small wave numbers k the 'sinc' term vanishes, i.e.

$$\frac{\sin(k\Delta x)}{k\Delta x} \to 0.$$

For large wave numbers, i.e. small wavelengths and thus small scales, the error becomes large. If a wavelength is approximated by n points, the wave number can be expressed as:

$$k = \frac{2\pi}{n\Delta x}.$$

Table 4.2 Overview of representative direct numerical simulations

Type	Re	$\mathcal{N}$	Reference
Isotropic turbulence	150	240^3	Vincent and Meneguzzi (1991)
Isotropic turbulence	1,200	4096^3	Ishihara et al. (2009)
channel flow	3,300	$192 \times 129 \times 160$	Kim et al. (1987)
Channel flow	34,200	$768 \times 768 \times 769$	del Álamo et al. (2004)
Boundary layer	1,410	$432 \times 80 \times 320$	Spalart (1988)
Boundary layer	4,300	$4096 \times 385 \times 384$	Schlatter et al. (2010)
Pipe flow	5,300	$96 \times 128 \times 256$	Eggels et al. (1994)
Pipe flow	44,000	$300 \times 1024 \times 2048$	Wu and Moin (2008)
Jet flow	2,400	$450 \times 80 \times 64$	Boersma et al. (1998)

For the differentiation of a simple wave with an error less than 5 % already 12 points are needed. Higher-order methods (i.e., methods that also use: $f(x \pm 2\Delta x)$, $f(x \pm 3\Delta x)$, ...) need in general fewer points for the same accuracy. The same is true for methods using *Chebyshev* or *Fourier* expansions. From the previous analysis it can be observed that numerical methods in general accurately predict the large scales and that the accuracy of the small scales is certainly an issue. This is especially the case in problems where the small scale play an important role, e.g. turbulent mixing and combustion.

The equations are in general advanced in time with an explicit numerical method for the non-linear terms. Popular methods are *Runge–Kutta* and *Adams–Bashford* methods. The viscous term is often integrated with an implicit *Crank–Nicolson* method. A fully implicit treatment is in general not advisable, because in that case the non-linear term has to be linearized, e.g. the mechanism responsible for turbulence generation is slightly modified (Table 4.2).

One of the first successful attempts of a DNS of turbulent flow is described by Kim et al. (1987), who simulated a turbulent channel flow at Reynolds number of 3,300 using about 4×10^6 grid points. The first DNS of turbulent pipe flow at $Re = 5,300$ was performed by Eggels et al. (1994); with the increase of computing power, contemporary simulations can represent a turbulent pipe flow with a Reynolds number of 44×10^3 (Wu and Moin 2008).[5] Despite such substantial increase in computer power, the Reynolds numbers that can be represented by such simulations are still far away from those of flows of industrial interest. For example, the Reynolds number of a commercial airliner at cruise speed is of the order of 10^9; see Problem 1 below.

Hence, it is evident that for many practical problems one has to resort to other approaches for the design and optimization of apparatus that involve turbulent flows. This implies that we must reduce the complexity of the equations of motion and make

---

[5]This increase in DNS performance that is reflected in the resolved flow Reynolds number from 1994 to 2008 is in accordance with *Moore's law* that computer resources double about every 18 months.

use of models that represent the effects of turbulence. This is the main topic of the remainder of this book. Nonetheless, DNS has provided substantial insights in the physical aspects of turbulence, and many of the processes that have been studied at low to moderate Reynolds numbers help us to understand what happens in practical situations at much higher Reynolds numbers. Therefore, DNS remains one of the principal tools for the fundamental investigation of turbulent flows.

Problem

1. Derive the expression in (4.22). Make an estimate of the required computational power to represent the flow around a commercial airliner at cruise speed in a direct numerical simulation. What is the required memory for the computation, and how long would it take to perform the simulation per integral time scale using a 1 PFLOPS ($=10^{15}$ *floating point operations per second*) computer? And what if you would use 1,000 of such computers simultaneously?

Chapter 5
Statistical Description of Turbulence

A turbulent flow is a chaotic and fluctuating flow state, and for most applications we are not interested in all the details of the flow. It thus seems obvious to focus our attention on statistical quantities, such as the average and the standard deviation of fluctuating flow variables. We first deal with the general requirements these statistical variables have to meet. Next we deduce the equations of motion for these variables.

5.1 Statistics

As a starting point we say that, in turbulent flow, an *instantaneous* quantity can be split into an *average* plus a *fluctuation*. Take for example the velocity component u_i, which can be expressed as

$$u_i = \overline{u_i} + u_i'. \tag{5.1}$$

The averaging operator is denoted by a bar above the quantity and the fluctuation is denoted by a prime. This expression (5.1) is commonly referred to as the *Reynolds decomposition*. If we want to be able to apply this averaging operator later in the equations of motion, it is required that this operator satisfies the following general conditions that are better known as the *Reynolds conditions*:

$$
\begin{aligned}
&(i) \quad \overline{f+g} = \overline{f} + \overline{g}, \\
&(ii) \quad \overline{\alpha f} = \alpha \overline{f}, \\
&(iii) \quad \overline{\dfrac{\partial f}{\partial s}} = \dfrac{\partial \overline{f}}{\partial s}, \\
&(iv) \quad \overline{\overline{f} g} = \overline{f}\,\overline{g}.
\end{aligned}
\tag{5.2}
$$

© Springer International Publishing Switzerland 2016
F.T.M. Nieuwstadt et al., *Turbulence*, DOI 10.1007/978-3-319-31599-7_5

Here f and g are fluctuating quantities and α is a constant. On the basis of these Reynolds conditions we can for example infer the following relationships:

$$\overline{u'} = 0, \quad \overline{\overline{u}} = \overline{u}, \quad \text{and} \quad \overline{\overline{f}\,\overline{g}} = \overline{f}\,\overline{g}. \tag{5.3}$$

How do we define the average? In practice, as we would do in a laboratory measurement, we often use the *time average*, which is defined as

$$\overline{u}^T (t) = \frac{1}{T} \int\limits_{-\frac{1}{2}T}^{+\frac{1}{2}T} u(t + \tau)\,d\tau, \tag{5.4}$$

where T represents the *averaging time*. In the same way we can define a *line average* as

$$\overline{u}^L (x) = \frac{1}{L} \int\limits_{-\frac{1}{2}L}^{+\frac{1}{2}L} u(x + \xi)\,d\xi. \tag{5.5}$$

For now we use superscripts to indicate what kind of average we are considering. It is not hard to see that we can generalize (5.5) to surface averages or space averages as well.

While the averaging operators defined in (5.4) and (5.5) are very useful in a practical sense, they have one big disadvantage: they do not meet the Reynolds conditions in (5.2); in particular they do not satisfy condition (iv). The reason for this is that when, for example, we apply expression (5.4) twice to compute $\overline{\overline{u}}$, the integration interval over which the function $u(t + \tau)$ should be known would double from T to $2T$. In other words, $\overline{\overline{u}} \neq \overline{u}$. For this reason, the definitions are theoretically not applicable. That is why we focus on another average: the so-called *ensemble average*.

The principle underlying the ensemble average is the following. Suppose that we repeat the turbulence experiment (with the same initial and boundary conditions) N times. Every *realization* of the experiment will in principle be different. The ensemble average is then, by definition, equal to

$$\overline{u} = \lim_{N \to \infty} \frac{1}{N} \sum_{\alpha=1}^{N} u^{(\alpha)}, \tag{5.6}$$

where the index α indicates the realization of the experiment. The definition in (5.6) meets all four Reynolds conditions in (5.2). In the following we therefore always interpret a bar above a quantity as the ensemble average.

For a mathematical description of the ensemble average it is most appropriate to use the theory of *stochastic processes*. In this theory variables are described in terms of a probability distribution. Take for example the turbulent velocity component u. We now introduce a probability distribution $p(u)$. The probability that $u^* < u < (u^* + du^*)$ is then equal to $p_M(u^*)du^*$. The superscript $*$ indicates a reference to a particular value of u. The subscript M means that a probability distribution is basically a function of the position $M(\underline{x}, t)$ in space-time. By integration over all values of u, we can rewrite (5.6) as

$$\overline{u}(\underline{x}, t) = \int_{\forall u} u^* \, p_M(u^*) \, du^*. \tag{5.7}$$

This expression is the formal definition of the ensemble average.

Using the probability distribution $p_M(u)$ we can calculate other statistics as well. An example is the *variance*, which is defined as

$$\overline{u'^2}(\underline{x}, t) = \int_{\forall u} (u^* - \overline{u})^2 \, p_M(u^*) \, du^*. \tag{5.8}$$

The *standard deviation* is then, by definition, equal to

$$\sigma_u(\underline{x}, t) = \sqrt{\overline{u'^2}}. \tag{5.9}$$

We can generalize this expression to

$$\overline{f(u)} = \int_{\forall u} f(u^*) \, p_M(u^*) \, du^*. \tag{5.10}$$

This description can easily be extended to two variables: for example, to the velocity components u_i and u_j. The probability that $u_i^* < u_i < u_i^* + du_i^*$ and that $u_j^* < u_j < u_j^* + du_j^*$ is now by definition equal to $p_M(u_i^*, u_j^*)du_i^* du_j^*$. With this it follows that

$$\overline{u_i u_j}(\underline{x}, t) = \iint_{\forall u_i, u_j} u_i^* u_j^* \, p_M(u_i^*, u_j^*) \, du_i^* \, du_j^*, \tag{5.11}$$

which is called the *covariance* at $M = (\underline{x}, t)$. It will be clear that this description can be extended to all other quantities that are defined at M, such as pressure and temperature.

Thus far, we have limited ourselves to the probability distribution p_M at a single position M. We call these *single-point distributions*. The quantities calculated with (5.7) and (5.8) using this method are called first-order and second-order *single-point moments*, respectively. The step towards stochastic variables defined at two points is now obvious. The probability that the variables u_i at $M_1 = M(\underline{x}_1, t_1)$ and u_j at $M_2 = M(\underline{x}_2, t_2)$ satisfy $u_i^* < u_i < u_i^* + du_i^*$ and $u_j^* < u_j < u_j^* + du_j^*$ equals $p_{M_1, M_2}(u_i^*, u_j^*)du_i^* du_j^*$. From this it follows that

$$\overline{u_i(\underline{x}_1, t_1) u_j(\underline{x}_2, t_1)} = \iint_{\forall u_i, u_j} u_i^* u_j^* \, \mathsf{p}_{M_1 M_2}(u_i^*, u_j^*) \, du_i^* \, du_j^*. \qquad (5.12)$$

This is called the correlation between u_i and u_j at M_1 and M_2, respectively, which is a *two-point average*. With this type of average we can acquire information on the *structure* of turbulence. We further elaborate on this in Chap. 9.

It can be proven that, for a complete statistical description of a turbulent field, we would have to specify the probability distribution at an infinite number of positions and for an infinite number of statistical moments. However, in this book we limit ourselves in most cases to single-point and two-point averages.

5.2   Stationarity and Homogeneity

We now consider a couple of special *stochastic* processes. First of all, we take a *stationary* process. The formal definition for such a process is that the probability distribution p is invariant to translations in time. For a single-point distribution this means that

$$\mathsf{p}_t = \mathsf{p}_{t+dt}. \qquad (5.13)$$

In other words, p is independent of t. Consequently, all single-point averages are independent of time: for example, $\overline{u}$ and $\overline{u'^2} \neq f(t)$. For two-point averages (for example at two times t_1 and t_2) stationarity would mean that the probability distribution is only a function of the *time difference* $t_1 - t_2$. Thus, for a stationary process it holds that for the correlation:

$$\overline{u_i(t_1) u_j(t_2)} = f(t_1 - t_2). \qquad (5.14)$$

The second special process is a *homogeneous* stochastic process. For this the probability distribution is invariant to translations along a line in space. In this case the single-point averages are independent of the coordinates along that line, and the two-point averages are just a function of the distance between the two points.

Until now we discussed several types of averages. On the one hand the time and spatial averages $\overline{u}^T$ and $\overline{u}^L$, respectively, and on the other hand the ensemble average $\overline{u}$. How can we relate these averages to each other? For this we invoke the so-called '*ergodicity hypothesis*.' On the basis of this hypothesis, it is assumed that for a *stationary* process

$$\lim_{T \to \infty} \overline{u}^T = \overline{u}. \qquad (5.15)$$

So, for a stationary process we can treat the time average as the ensemble average. On the basis of the same theorem, it holds that for a *homogeneous* process that

$$\lim_{L \to \infty} \overline{u}^L = \overline{u}. \qquad (5.16)$$

For details on the proofs for (5.15) and (5.16) we refer to the literature and to one of the problems at the end of this section.

Qualitatively, we can understand these results. In a stationary process the fluctuations have by definition a limited time scale, which we later identify with the *integral time scale* $\mathcal{T}$. When we now consider an averaging time that is sufficiently long, i.e. $T > \mathcal{T}$, we can think about it as constructed from many intervals with a duration $\mathcal{T}$, which are statistically independent. It now follows that the time average must be equivalent to the ensemble average.

In Fig. 5.1 is shown the instantaneous velocity field in a planar cross section of a turbulent pipe flow (measured with PIV), for which the axial (x) direction is homogeneous. The mean velocity profile $\bar{u}(r)$ was determined from an ensemble of 100 statistically independent measurements. Subtraction of the mean velocity profile yields the fluctuating velocity field (u', v'). This clearly reveals the turbulent flow structures near the wall and in the central region of the pipe.

Problem

1. The mean quadratic difference between the time average and the ensemble average reads:

$$I = \overline{(\bar{u} - \bar{u}^T)^2}.$$

Assuming stationarity, show that the following holds:

$$I = \frac{2}{T} \int_0^T \left(1 - \frac{\tau}{T}\right) R(\tau)\, d\tau,$$

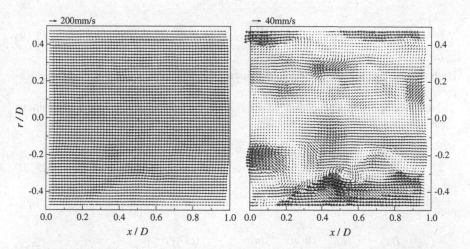

Fig. 5.1 Instantaneous flow field in a planar cross section through the centerline of a turbulent pipe flow at $Re = 5{,}300$. The *top* and *bottom* of the graph coincide with the inner pipe wall. The *left arrow* plot shows the absolute velocities (u, v), while the *right arrow* plot shows the fluctuations (u', v') after subtraction of the mean velocity profile $\bar{u}(r)$, revealing the instantaneous turbulent flow structure. (The *arrows* in the *right* plot have been enlarged 6 times with respect to those in the *left* plot.) Data from: Westerweel et al. (1996)

where $R(t'' - t') = \overline{u(t')u(t'')}$ is the time correlation function. Derive on the basis of this equation an estimate of the error in the time average.

5.3 The Reynolds Equations

The decomposed instantaneous quantities, which consist of an average and a fluctuation, are now applied to the equations of motion. For this we define

$$u_i = \overline{u_i} + u_i'$$
$$p = \overline{p} + p' \qquad\qquad (5.17)$$
$$\theta = \overline{\theta} + \theta'$$

where we emphasize again that the averaging operator is actually the ensemble average.

The procedure is now as follows. We substitute (5.17) into the equations of motion for the instantaneous quantities. The result then becomes an average, and this yields the equations for the average flow. After this we subtract the equations for the average flow from the original equations of motion, yielding the equation for the fluctuating quantities.

Application of this method to the continuity equation (2.20) for incompressible flows yields

$$\frac{\partial \overline{u_i}}{\partial x_i} = 0, \qquad\qquad (5.18)$$

$$\frac{\partial u_i'}{\partial x_i} = 0. \qquad\qquad (5.19)$$

Both the average velocity field and the fluctuating velocity field are thus *divergence-free*.

Next, we apply this procedure to the Navier–Stokes equations in the Boussinesq approximation (2.21) with the result

$$\frac{\partial \overline{u_i}}{\partial t} + \frac{\partial \overline{u_i}\,\overline{u_j}}{\partial x_j} = -\frac{1}{\rho_0}\frac{\partial \overline{p}}{\partial x_i} + \frac{g}{T_0}\overline{\theta}\delta_{i3} + \nu\frac{\partial^2 \overline{u_i}}{\partial x_j^2} - \frac{\partial \overline{u_i' u_j'}}{\partial x_j}, \qquad\qquad (5.20)$$

and:

$$\frac{\partial u_i'}{\partial t} + \overline{u_j}\frac{\partial u_i'}{\partial x_j} + u_j'\frac{\partial \overline{u_i}}{\partial x_j} + \frac{\partial u_i' u_j'}{\partial x_j} - \frac{\partial \overline{u_i' u_j'}}{\partial x_j} = -\frac{1}{\rho_0}\frac{\partial p'}{\partial x_i} + \frac{g}{T_0}\theta'\delta_{i3} + \nu\frac{\partial^2 u_i'}{\partial x_j^2}. \qquad (5.21)$$

For the energy conservation equation (2.22) we find

$$\frac{\partial \overline{\theta}}{\partial t} + \overline{u}_j \frac{\partial \overline{\theta}}{\partial x_j} = \kappa \frac{\partial^2 \overline{\theta}}{\partial x_j^2} - \frac{\partial \overline{u'_j \theta'}}{\partial x_j} \tag{5.22}$$

and

$$\frac{\partial \theta'}{\partial t} + \overline{u}_j \frac{\partial \theta'}{\partial x_j} + u'_j \frac{\partial \overline{\theta}}{\partial x_j} + \frac{\partial \overline{u'_j \theta'}}{\partial x_j} - \frac{\partial \overline{u'_j \theta'}}{\partial x_j} = \kappa \frac{\partial^2 \theta'}{\partial x_j^2}. \tag{5.23}$$

In Eqs. (5.20) and (5.22) the *molecular terms* are negligible. This is because the average quantities in these equations scale with the macroscopic scales. The ratio of the inertia terms compared to the molecular terms is then proportional to the *Reynolds number*, and we know this Reynolds number is large for a turbulent flow.

Apart from this, we observe that in these equations new and previously unfamiliar terms appear for $\overline{u}_i$ and $\overline{\theta}$: $\overline{u'_i u'_j}$ and $\overline{u'_j \theta'}$. These terms are called the *Reynolds terms*.

We first focus on the term $\overline{u'_i u'_j}$. This term can be interpreted as the transport in the j-direction of momentum per unit mass in the i-direction. Such a transport of momentum has the same effect as a stress acting on a surface, and for this reason it is commonly referred to as the *Reynolds stress*. Figure 5.2 shows a measurement of the stream wise and wall-normal velocity components in a turbulent flow near a solid wall. The velocity fluctuations u' and v' are obviously correlated, yielding a *negative* Reynolds stress $\overline{u'v'}$; this implies a momentum transfer in which slower moving fluid is transported away from the wall and faster moving fluid towards the wall, i.e. a net transfer of momentum towards the wall that induces a (friction) force on the plate. In a similar way we can interpret the term $\overline{u'_j \theta'}$ as temperature transport.

Both transport terms $\overline{u'_i u'_j}$ and $\overline{u'_j \theta'}$ originate from the nonlinear advection terms in the Navier–Stokes equations and the energy equation.

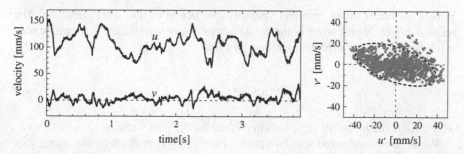

Fig. 5.2 (*left*) Time series of the measured stream wise (u) and wall-normal (v) velocity (measured by means of laser-Doppler velocimetry) in a turbulent boundary layer over a flat plate (see Sect. 6.2). (*right*) The correlation of the fluctuations $u' = u - \overline{u}$ and $v' = v - \overline{v}$. The ellipse represents the correlation between the two velocity components, i.e. $\overline{u'v'}$, or *Reynolds stress*. The main axes of the ellipse are proportional to σ_u and σ_v, respectively. The measurement was taken at a distance of $y^+ = 20$ viscous wall-units from the plate; see (6.19) for a definition of wall units (Data courtesy of A.D. Schwarz-van Manen.)

The problem is that both terms are new unknown quantities in the equations for $\overline{u}_i$ and $\overline{\theta}$, and because of this the number of unknown quantities increases past the number of equations. We thus need another relationship to solve our problem. This is known as the *closure problem*. This is one of the central problems in theory of turbulence.

To get to a first closure hypothesis, we compare the *turbulent stress tensor* $\Sigma_{ij} = -\rho\overline{u'_i u'_j}$ to the *molecular stress tensor* $\sigma_{ij} = -p\delta_{ij} + \tau_{ij}$, which we encountered before in Sect. 2.1. The first term in σ_{ij} is the isotropic *pressure* and the second term is known as the *deviatoric stress tensor*. In other words, this is the stress that relates to deviations from isotropy. For a Newtonian fluid the deviatoric stress is directly proportional to the *strain rate tensor* s_{ij}, with the result

$$\tau_{ij} = 2\mu s_{ij} = \rho_0 \nu \left(\frac{\partial u_i}{\partial x_j} + \frac{\partial u_j}{\partial x_i} \right). \tag{5.24}$$

This equation is called the *constitutive equation*.

Analogously, we write for the turbulent stress tensor Σ_{ij}:

$$\Sigma_{ij} = -\frac{1}{3}\rho_0 \overline{u_k'^2} \delta_{ij} + \rho_0 \left(-\overline{u'_i u'_j} + \frac{1}{3}\overline{u_k'^2}\delta_{ij} \right). \tag{5.25}$$

Here the first term is referred to as the *normal stresses*, which we recognize as equivalent to a 'turbulent' *pressure*. However, in most flows the turbulent pressure is negligible compared to the static pressure p. The second term on the right-hand side of (5.25) can be considered as the deviatoric strain. On the basis of (5.24) it is obvious to write this as

$$\rho_0 \left(-\overline{u'_i u'_j} + \frac{1}{3}\overline{u_k'^2}\delta_{ij} \right) = 2\rho_0 K \, \overline{s_{ij}} = \rho_0 K \left(\frac{\partial \overline{u_i}}{\partial x_j} + \frac{\partial \overline{u_j}}{\partial x_i} \right). \tag{5.26}$$

Here K is the so-called turbulent viscosity coefficient or the '*eddy viscosity*.' The relationship (5.26) is known as the *Boussinesq closure hypothesis*. In a similar way, it follows for the turbulent temperature flux that

$$-\overline{u'_i \theta'} = K_\theta \frac{\partial \overline{\theta}}{\partial x_i}, \tag{5.27}$$

where K_θ is the *turbulent diffusion coefficient* for the heat transfer.

We are now confronted with two problems. First, under which circumstances are an hypotheses like (5.26) and (5.27) valid, and second, how do we determine K and K_θ? We deal with these problems in the following section.

Problems

1. Under what circumstances do you expect that the turbulent pressure becomes equal the static pressure?

5.4 Kinetic Theory of Momentum Transfer

To understand the transport of momentum we first focus on the molecular stress tensor, and we ask ourselves the following question: How can we couple the friction forces in a flow to the molecular structure? Let us limit ourselves to an *ideal gas*, where momentum is only transferred through collisions between individual molecules. To this model gas we apply a simple kinetic theory to describe the momentum transfer.

Consider the special case of a *one-dimensional flow* with a velocity $U(y)$ as illustrated in Fig. 5.3. According to the continuum hypothesis we have to interpret the velocity $U(y)$ as the average of the velocities of a large number of individual molecules in a small volume. Now consider a single molecule at the level y_0, and thus with an average velocity in the x-direction equal to $U(y_0)$. The velocities of the individual molecules of course deviate from this average. Assume here that the molecule has a velocity component in the direction of the y-axis equal to a, where a represents the characteristic velocity of individual molecules. This characteristic velocity is by definition equal to the *speed of sound*, since that is, in a matter of sense, the speed at which information can propagate through the fluid.

The molecule displaces itself over a vertical distance λ (where λ is called the *mean free path*) before it collides with a molecule at the level $y_0 - \lambda$. This last molecule has an average horizontal velocity $U(y_0 - \lambda)$.

Due to this collision, the horizontal momentum M of the gas at the level $y_0 - \lambda$ changes by an amount of

$$\Delta M = I_1 - I_2 = m \{U(y_0) - U(y_0 - \lambda)\}, \qquad (5.28)$$

where m is the mass of the molecule. Per unit time and per unit area there are N_t collisions, which is

$$N_t \approx a\,n, \qquad (5.29)$$

where n is the number of molecules per unit volume. The total change in momentum per unit time and unit area now follows from the product of (5.28) and (5.29). According to Newton's laws, we can interpret this as a *force per unit area*, that is a *stress*, which is

Fig. 5.3 Kinetic momentum transfer by collisions of molecules in a gas for one-dimensional shear flow $U(y)$. The molecules move at approximately the speed of sound a with a mean-free path λ

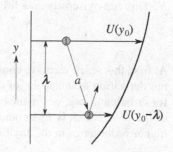

$$\tau_{12} \approx N_t \Delta M = \alpha \, n \, m \, a \left\{ U(y_0) - U(y_0 - \lambda) \right\}, \tag{5.30}$$

where α is a proportionality constant that equals about $2/3$. The term τ_{12} is the stress in the x-direction that is applied by the gas at position y_0 to the gas at position $y_0 - \lambda$, which is called the *shear stress*. Using a Taylor series expansion of Eq. (5.30), it follows that

$$\tau_{12} = \alpha \rho a \lambda \frac{\partial U}{\partial y} \left(1 + \frac{1}{2} \frac{\lambda}{L} + \cdots \right), \tag{5.31}$$

where $\rho = nm$ is the density of the gas, and L represents a characteristic length scale, defined as

$$L = \frac{\partial U}{\partial y} \Big/ \frac{\partial^2 U}{\partial y^2}. \tag{5.32}$$

In all cases for which the continuum hypothesis is valid, $\lambda \ll L$, so that

$$\tau_{12} = \rho \nu \frac{\partial U}{\partial y}, \tag{5.33}$$

where $\nu = \alpha a \lambda$ is the kinematic viscosity. This expression is consistent with (5.24). With $\lambda \approx 7 \times 10^{-8}$ m and $a \approx 3.4 \times 10^2$ m/s, it follows that: $\nu \approx 1.5 \times 10^{-5}$ m²/s, which is a representative value for the kinematic viscosity of air.

We thus find that our hypothesis (5.24) for a Newtonian fluid can in this case be supported by a molecular model. Next we attempt to apply this derivation to the turbulent shear stress $-\overline{u_1' u_2'}$ in a one-dimensional flow $\overline{u}(y)$. For this we apply the following substitution to the argumentation above:

$$\lambda \Rightarrow \mathcal{L}$$
$$a \Rightarrow \mathcal{U}$$

where $\mathcal{L}$ and $\mathcal{U}$ are characteristic length and velocity scales of an eddy.

Instead of colliding molecules, we now consider interacting 'eddies.' After an analysis that is similar to the analysis above, we find an expression for the turbulent shear stress $-\overline{u_1' u_2'}$ that similar to (5.33), where ν is replaced by the eddy viscosity K. This eddy viscosity now follows from

$$K \sim \mathcal{U} \mathcal{L} \tag{5.34}$$

At first, this result seems to confirm the Boussinesq closure hypothesis. However, in our derivation we made use of the assumption $\mathcal{L} \ll L$, and this is not generally true for turbulent flows, where the characteristic scale $\mathcal{L}$ is of the order of the geometry of the flow, which is L. In other words, we cannot disregard the second-order and higher-order terms in the Taylor series.

From this we have to conclude that the Boussinesq closure hypothesis, sometimes referred to as K-theory, is, to say the least, only a crude approximation of reality. This theory suggests that turbulent stresses at certain positions can be described using local quantities only (such as the strain rate tensor). This goes well for molecular stresses, but turbulence is not 'local,' because $\mathcal{L} \sim L$. In short, K-theory is completely lacking a physical basis. Consequently, the application of K-theory is purely empirical.

Yet, in practice it turns out that K-theory can provide an apt approximation of reality under quite a number of circumstances. This holds in particular for flows with simple geometries. That is why for now we maintain to apply K-theory, or the Boussinesq closure hypothesis. We have to be aware, however, of the limitations of this theory. At a later stage we look at some more complicated closure models.

We have found that the eddy viscosity is proportional to the characteristic velocity and length scale of the turbulence. This is in agreement with our phenomenological findings in Sect. 4.2. However, our closure problem is not yet completely solved, because the actual value of K is not yet determined. We will thus have to specify a relationship describing K in terms of known quantities. These are often empirical relationships. The most common one is the so-called *Prandtl mixing length hypothesis*. This hypothesis is based on an estimate of $\mathcal{U}$ according to

$$\mathcal{U} \sim \mathcal{L} \left| \frac{\partial \overline{u}}{\partial y} \right|,$$

and with this it follows that

$$K = \mathcal{U}\mathcal{L} = \mathcal{L}^2 \left| \frac{\partial \overline{u}}{\partial y} \right|. \tag{5.35}$$

The *mixing length* $\mathcal{L}$ ('*Mischungsweg*') is however still unknown, and for every flow geometry the mixing length differs. This again underlines the empirical character of K-theory.

Von Kármán proposed a formulation for K that no longer contains any unknown quantities. It reads

$$K = kL^2 \left| \frac{\partial \overline{u}}{\partial y} \right|, \tag{5.36}$$

where L is given by (5.32) and where k is the *von-Kármán constant* with a value $k \approx 0.4$. This formulation is no longer generally used, since it only yields correct results for turbulent flows in the vicinity of a wall (as we will see in the following chapter). In all other turbulent flows, Eq. (5.36) fails completely.

Problems

1. The smallest flow scale in a turbulent fluid is the Kolmogorov scale. For the continuum hypothesis to hold in a turbulent fluid, the condition $\lambda \ll \eta$ should be met, where λ is the mean free path. Show that this condition can be written as $Re^{1/4}Ma^{-1} \gg 1$, where Ma represents the *Mach number*: $Ma = u/a$.

Chapter 6
Turbulent Flows

In this chapter we describe some basic turbulent flows in detail, where the turbulence is generated by *shear*. These flows can be divided into two categories.

The first category consists of so-called *wall turbulence*. These are flows near a fixed wall, such as channel flow (see Sect. 6.1), pipe flow, and a turbulent boundary layer over a flat wall. The other category entails so-called *free turbulence*. These are flows that are unaffected by a wall, and thus take place, so to speak, in free space. An example is the *turbulent jet* (see Sect. 6.6). Other free turbulent flows are the turbulent wake behind a bluff body and the turbulent mixing layer.

6.1 Channel Flow

In this section we consider the flow in a channel with parallel walls, as illustrated in Fig. 6.1. Our point of departure in the analysis of this flow is a stationary and horizontally homogeneous turbulent flow. In other words, for the average quantities it holds that: $\partial/\partial t = 0$, and $\partial/\partial x = \partial/\partial z = 0$ (for the velocity only). The average velocity is given by: $\overline{u_i} = (\overline{u}, \overline{v}, 0)$, that is, $\overline{u} = 0$. Variations in fluid density and temperature are not taken into consideration. For this geometry the equations of motion can be simplified substantially.

Given that for the velocity $\partial/\partial x = \partial/\partial z = 0$, the Reynolds-averaged continuity equation (5.18) reduces to

$$\frac{\partial \overline{v}}{\partial y} = 0, \tag{6.1}$$

and using the boundary condition $\overline{v} = 0$ for $y = 0$, it follows that $\overline{v} \equiv 0$ ($\forall y$), or, in other words, the average velocity has a component in the x-direction only, so that: $\overline{u_i} = (\overline{u}(y), 0, 0)$.

© Springer International Publishing Switzerland 2016
F.T.M. Nieuwstadt et al., *Turbulence*, DOI 10.1007/978-3-319-31599-7_6

Fig. 6.1 The geometry of a channel flow between two parallel plates separated by a distance $2H$

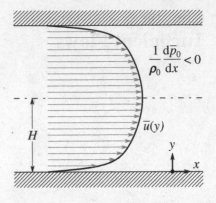

Again, given that: $\partial/\partial x = \partial/\partial z = 0$, and $\partial/\partial t = 0$, the Reynolds-averaged Navier-Stokes equations (5.20) in the x- and y-directions reduce to:

$$0 = -\frac{\partial \overline{p}}{\partial x} + \frac{\partial \tau_t}{\partial y}, \tag{6.2}$$

with $\tau_t/\rho_0 = -\overline{u'v'} + \nu\, \partial\overline{u}/\partial y$, and:

$$0 = -\frac{\partial \overline{p}}{\partial y} - \rho_0 \frac{\partial \overline{v'^2}}{\partial y}. \tag{6.3}$$

This last equation yields for the pressure: $\overline{p}/\rho_0 = f(x) - \overline{v'^2}$, which means that: $\partial\overline{p}/\partial x \neq f(y)$. Hence, the pressure in the channel can be written as:

$$\overline{p}(x, y) = \overline{p}_0(x) - \rho_0\overline{v'^2}(y), \tag{6.4}$$

where $\overline{p}_0$ is the mean pressure at the channel wall, which is evidently a function of x only. Hence, $\partial\overline{p}/\partial x = d\overline{p}_0/dx$.

The boundary conditions for this flow read:

$$\tau_t = 0 \quad \text{for} \quad y = H, \quad \text{and:} \quad \overline{u} = 0 \quad \text{for} \quad y = 0. \tag{6.5}$$

Next, we introduce the *wall friction velocity* u_*, which is defined as:

$$u_* = \sqrt{\frac{\tau_s}{\rho_0}}, \tag{6.6}$$

where τ_s is the *wall shear stress*. It thus holds that:

$$\tau_s = \rho_0 u_*^2 = \tau_t(0). \tag{6.7}$$

A relation exists between u_* and $d\overline{p}/dx$. We can find this by integrating (6.2) in the y-direction between 0 and H, with the result:

$$u_*^2 = -\frac{H}{\rho_0}\frac{d\overline{p}_0}{dx}. \tag{6.8}$$

With this we can write the integral of (6.2) as:

$$\frac{\tau_t}{\rho_0} \equiv -\overline{u'v'} + \nu\frac{\partial\overline{u}}{\partial y} = u_*^2\left(1 - \frac{y}{H}\right). \tag{6.9}$$

In other words, the *total* shear stress is a *linear* function of y.

Until now all results have been exact. However, without introducing a closure hypothesis we can not proceed any further. This subsequent step in the analysis is the focus of the next section.

6.2 Mean Velocity Profile

In older literature, the mean velocity profile in a turbulent channel or pipe flow is commonly represented as a $1/n$th power law with $n = 7$ for intermediate Reynolds numbers of 10^4–10^5 (Schlichting 1979; White 2011). This form is purely empirical and lacks any physical background. In Fig. 6.2 such a profile is compared against experimental data. Although the agreement is reasonable in the central region of the flow, significant deviations occur near the wall. In addition, the power-law velocity gradient is discontinuous at the centerline, and it diverges at the wall, i.e., $|\partial u/\partial y| \to \infty$ for $y \to 0$ (implying infinite friction!). Instead, the mean velocity profile derived in this section is founded on the equations of motion, in combination with physically acceptable closure models.

In order to compute the velocity profile in a turbulent channel flow, based on the equations of motion derived in the previous section, we use of the Prandtl mixing length hypothesis, defined in (5.35). In this hypothesis an unknown mixing length occurs, denoted by $\mathcal{L}$. We can also interpret this length $\mathcal{L}$ as a characteristic length scale of an eddy. The choice of $\mathcal{L}$ is dependent on the location in the channel. Below we consider four regions where we use different definitions for $\mathcal{L}$.

I. Core Region

We first consider the region in the center (or the core) of the channel. In that case the characteristic size of the eddies scales with the dimension of the channel, that is:

$$\mathcal{L} = \beta H. \tag{6.10}$$

Recall that in Sect. 4.2 the point of departure for the analysis was that the characteristic size of the eddies scales with the geometry of the flow domain. In the core region

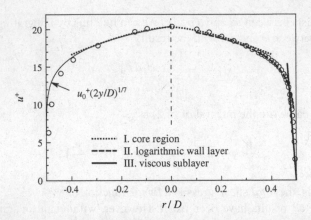

Fig. 6.2 The dimensionless mean velocity profile u^+ of a turbulent flow in a pipe at $Re = 10,000$ as a function of the distance r from the centerline of a pipe with diameter D, compared to experimental data from den Toonder and Nieuwstadt (1997). On the *right half* are plotted the velocity profiles in (6.12), (6.15), and (6.18). On the *left half* is plotted a $\frac{1}{7}$th power profile (Schlichting 1979)

of the channel the flow is completely turbulent, and viscous stresses are negligible compared to the Reynolds stress. The equation for the velocity profile in the core region then reduces to:

$$\beta^2 H^2 \left| \frac{\partial \overline{u}}{\partial y} \right| \frac{\partial \overline{u}}{\partial y} = u_*^2 \left(1 - \frac{y}{H} \right), \tag{6.11}$$

with the solution:

$$\overline{u} = u_0 - \frac{2}{3} \frac{u_*}{\beta} \left(1 - \frac{y}{H} \right)^{3/2}. \tag{6.12}$$

Here u_0 is an integration constant that represents the velocity at the center plane of the channel. The value of β can be obtained from experimental data. For turbulent pipe flow the value for β is about 0.13; see Fig. 6.2.

II. Wall Region

When we get closer to the wall, the eddies become limited in size because of the influence of the wall. This means that (6.10) is no longer valid, and it thus follows for $\mathcal{L}$:

$$\mathcal{L} = k\, y. \tag{6.13}$$

Here, k is the 'Von Kármán' constant, which was introduced previously in Sect. 5.4. On the basis of experiments a value of $k \cong 0.4$ has been found for this constant.

Moreover, close to the wall experimental data indicate that $\tau_t \cong \rho_0 u_*^2$. For this reason, the wall region is sometimes called the *constant stress layer* or *surface layer*. If we assume that the flow is completely turbulent, then the following equation holds:

$$k^2 y^2 \left| \frac{\partial \overline{u}}{\partial y} \right| \frac{\partial \overline{u}}{\partial y} = u_*^2,$$

(6.14)

with the solution:

$$\overline{u} = \frac{u_*}{k} \ln \left(\frac{y}{y_0} \right).$$

(6.15)

Here, y_0 is an integration constant. The velocity profile is thus *logarithmic*, and therefore this region is also referred to as the *logarithmic wall layer*. The result (6.15) is one of the most fundamental properties of the turbulent velocity profile near a solid wall. In a later section of this book we return to this velocity profile and to the interpretation of the constant y_0.

III. Viscous Sublayer

It is clear that (6.15) is not valid for $y = 0$. Thus, one can expect a different velocity profile very close to the wall. We first consider the *eddy viscosity K*, which in this case is defined by the relation: $u_*^2 = K \partial \overline{u} / \partial y$. With (6.15), it follows that

$$K = k u_* y.$$

(6.16)

We see that K becomes smaller and smaller as we approach the wall. At a certain distance from the wall, $K \cong \nu$. This means that the flow can no longer be turbulent, since viscosity effects become dominant. Turbulent stresses are negligible here, and for the equation of motion it follows that:

$$u_*^2 = \nu \frac{\partial \overline{u}}{\partial y},$$

(6.17)

with the solution:

$$\overline{u} = \frac{u_*^2}{\nu} y.$$

(6.18)

This expression satisfies the boundary condition: $\overline{u} = 0$ for $y = 0$.

The region where (6.18) is valid is called the *viscous sublayer*. This does not mean that the flow is laminar here, because that would imply that there are no fluctuations whatsoever. In the viscous sublayer velocity fluctuations do occur, but they are induced by the turbulence above the viscous sublayer.

We have now deduced the velocity profile in three separate regions. In order to determine the complete velocity profile, we need to specify the so-called *matching conditions*.

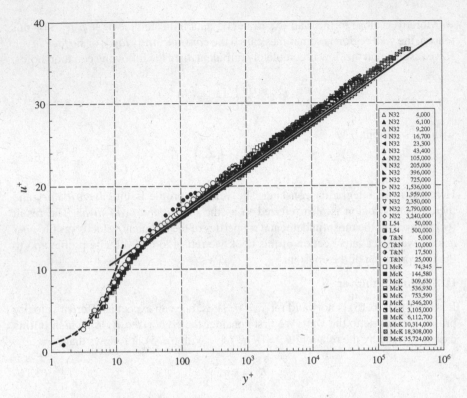

Fig. 6.3 Dimensionless mean velocity profile u^+ as a function of the dimensionless wall distance y^+ for turbulent pipe flow with Reynolds numbers between 4×10^3 and 36×10^6. Data from Nikuradse (1932), Laufer (1954), den Toonder and Nieuwstadt (1997), and McKeon et al. (2004). The *solid line* (——) corresponds to the logarithmic profile in (6.15); the *broken line* (− − −) corresponds to the linear profile in (6.18). The *open symbols* (o) correspond to the data in Fig. 6.2

IV. Viscous Sublayer ⇔ Logarithmic Layer

In order to match the viscous sublayer to the logarithmic layer, we introduce the dimensionless variables y^+ and u^+, which are defined as follows:

$$u^+ = \frac{\overline{u}}{u_*}, \quad \text{and:} \quad y^+ = y\frac{u_*}{\nu}. \tag{6.19}$$

These dimensionless units for wall distance and velocity are commonly referred to as *wall units*.

Figure 6.3 shows the results of several measurements of the mean velocity profile in the viscous sublayer and in the logarithmic layer. It follows that the viscous sublayer profile (6.18) is valid for $y^+ < 5$ and that the logarithmic profile (6.15) is valid for $y^+ > 30$. The value $y^+ = 5$ can be interpreted as the thickness δ_v of the viscous sublayer, or

$$\delta_v = 5\,\frac{\nu}{u_*}.$$

It is found that (6.15) and (6.18) are in good agreement, provided that the two solutions match at $y^+ = 11$ (see Fig. 6.3). This means that y_0 equals

$$y_0 = 0.135 \frac{\nu}{u_*}. \tag{6.20}$$

In other words, y_0 is proportional to the thickness of the viscous sublayer.

Thus, the mean velocity profile is given by:

$$u^+ = \begin{cases} y^+ & 0 < y^+ < 5 \\ \frac{1}{k} \left[\ln(y^+) + \Pi \right] & y^+ > 30 \end{cases} \tag{6.21}$$

where k is the *Von Kármán constant* with a value $k \cong 0.4$, and $\Pi = 2.0$, which follows from the intersection of the two solutions at $y^+ = 11$; see Fig. 6.3. Conventionally, we refer to the intermediate region $5 < y^+ < 30$ as the *buffer layer*, where both the viscous stress and the turbulent stress are important, but no simple solution exists. However, the mean velocity profile in the buffer layer and logarithmic layer can be described by a *displaced* log-law (Hunt et al. 2006), that is

$$u^+ = \frac{1}{k} \left[\ln \left(y^+ - \ell_v^+ \right) + \Pi \right], \tag{6.22}$$

with: $\ell_v^+ \cong 6$. The physical interpretation is that the strong shear layer that occurs at the top of the viscous sublayer effectively blocks eddies to reach the solid wall and implies a scaling of the eddy length scale in the turbulent flow above the viscous sublayer proportional to $(y - \ell_v)$.

V. Logarithmic Layer ⇔ Core Region

For the transition between the core region and the logarithmic wall layer, as described by (6.12) and (6.22) respectively, we assume that the mean velocity profiles are matched at a position $y = \alpha H$. From this it follows that:

$$\frac{u_*}{k} \left\{ \ln \left(\alpha \frac{u_* H}{\nu} \right) + 2.0 \right\} = u_0 - \frac{2}{3\beta} u_* (1 - \alpha)^{3/2}, \tag{6.23}$$

where the integration constant u_0, which is yet unknown, is expressed in other variables. Experimental data indicate that $\alpha \cong 0.15$.

Now on the basis of (6.12), (6.21) and (6.23) we have a mean velocity profile that is valid over the whole channel. We can consider the expression in (6.23) as an equation for u_0/u_*. We call this a *friction law*, which is further elaborated in Sect. 6.3.

In Figs. 6.4 and 6.5 are shown the turbulent stress and viscous stress measured in a turbulent pipe flow at $Re = 10,000$ (den Toonder and Nieuwstadt 1997). In the core region (I) the total stress is dominated by the turbulent stress and is directly proportional to the distance from the centerline; in the wall layer (II) the turbulent stress is nearly constant, while the viscous stress is still negligible; in the viscous

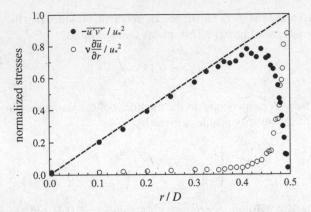

Fig. 6.4 The total normalized stress $\tau_t/\rho_0 u_*^2$ (— — —), defined in (6.9), as a function of the distance r from the centerline of a pipe with diameter D. The total stress consists of a contribution from the Reynolds stress $-\overline{u'v'}$ (•) and the viscous stress $\nu \partial \overline{u}/\partial r$ (○). Experimental data for a turbulent pipe flow at $Re = 10,000$ (den Toonder and Nieuwstadt 1997); cf. Fig. 6.2

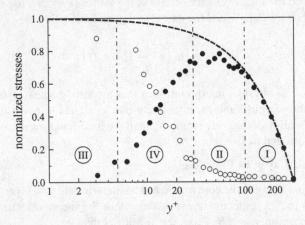

Fig. 6.5 The same data as in Fig. 6.4, but now as a function of the dimensionless distance y^+ from the pipe wall in a semi-log plot. *I* core region; *II* logarithmic wall region; *III* viscous sublayer; *IV* buffer layer. Experimental data for a turbulent pipe flow at $Re = 10,000$ with $D^+ = 624$ (den Toonder and Nieuwstadt 1997) (note that $r = 0$ corresponds to $y^+ = 312$ and $r = \frac{1}{2}D$ to $y^+ = 0$)

sublayer (III) the viscous stress is dominant, while in the buffer layer (IV) both the viscous stress and the turbulent stress are important. The logarithmic wall region ranges from $y^+ = 30$ to $r/D = 0.34$ ($y^+ \cong 100$), where the data begin to deviate from the profile in (6.22); see also Fig. 6.3. This corresponds to a value of $\alpha = 0.16$.

Problem

1. Consider a turbulent *Couette* flow, defined as the flow between two plane walls moving at a speed U_0 relative to each other. The distance between the walls is $2H$.

(a) Identify the different regions in this turbulent flow, and deduce for each of those regions an expression for its velocity profile.

(b) Deduce an expression for the friction coefficient $c_f = 2(u_*/U_0)^2$, see (6.31), when we use the Prandtl mixing length hypothesis for the eddy viscosity K in the core region. Repeat this calculation for a constant eddy viscosity K in the core region.

(c) Calculate or estimate all the terms in the kinetic energy equation (see Chap. 7) of the average flow and of the turbulence. Sketch the profiles of these terms.

6.3 Scaling of Turbulent Wall Flows

In the previous section we focused on the special geometry of channel flow. In this section these results are generalized to any *turbulent wall flow*. We mean by a wall flow basically any turbulent flow in the vicinity of a solid wall. Besides channel flow or pipe flow, another example of this type of turbulent flow is boundary layer flow, as illustrated in Fig. 6.6. Both channel flow and pipe flow are homogeneous in the axial direction, i.e., in the direction of the x-axis as defined in Fig. 6.1. However, boundary layer flow is not horizontally homogeneous, but develops as a function of the x-coordinate; see Fig. 6.6.

This development is slow, as can be seen in Fig. 6.7, where a visualization of a boundary layer over a flat plate is shown. We can describe such a slow development using the boundary layer equations that are discussed in Sect. 6.6. For now it can be assumed that this development does not directly influence the structure of the turbulence. The turbulent boundary layer is, so to speak, in a local equilibrium at every x-position. For this reason we can describe the turbulent boundary layer structure in the same way we could describe the turbulent flow in a channel or pipe.

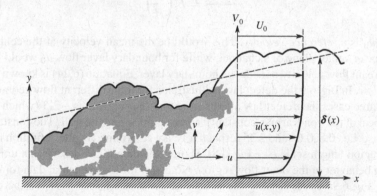

Fig. 6.6 Another wall-bounded turbulent flow: the turbulent boundary layer; cf. turbulent channel or pipe flow in Fig. 6.1. Illustration adapted from: Tennekes and Lumley (1972)

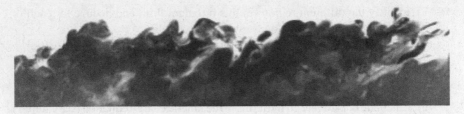

Fig. 6.7 Visualization of a turbulent boundary layer flow over a flat plate. Image from Van Dyke (1982)

Fig. 6.8 Scaling regions in turbulent wall flow

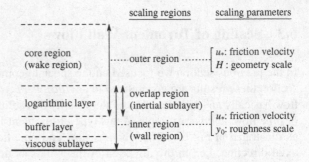

In every turbulent wall flow we can distinguish two scaling regions: the *outer region* and the *inner region*. Both regions can be described in terms of characteristic scaling parameters. This is illustrated in Fig. 6.8.

For the *outer region* the velocity scale is the friction velocity u_* and the length scale is H, where H is a characteristic length scale of the flow geometry: for example, the diameter of a pipe or the thickness of a boundary layer. On the basis of these observations we find that in the outer region the following scaling law is valid for the velocity profile:

$$\frac{u_0 - \overline{u}}{u_*} = F\left(\frac{y}{H}\right), \tag{6.24}$$

where u_0 is a *reference velocity*. This would be the mean velocity at the centerline of a pipe or center plane of a channel, while for boundary layer flow u_0 would be the free-stream flow velocity outside the boundary layer. Equation (6.24) is known as the *defect law*. In Fig. 6.9 the defect law is illustrated for three different flow geometries. In all three cases the defect law attains the same 'slope', that is -2.5 (which is the reciprocal of the Von Kármán constant), when approaching the wall. The constants in each case, i.e., 0.8, 0.65 and 2.35 respectively, are determined by the definition for the outer region length scale (see Fig. 6.8). Hence, the defect law describes a universal scaling behavior of the outer flow region. Note how the experimental data for defect law in Fig. 6.9a collapse onto the solid line towards smaller y/R for increasing Reynolds number. This implies that full universality of the outer scaling is reached for $Re \to \infty$.

For the *inner region* the scaling parameters are u_* and y_0. In the next section it is shown that y_0 is a length scale that relates to the properties of the wall. On the basis

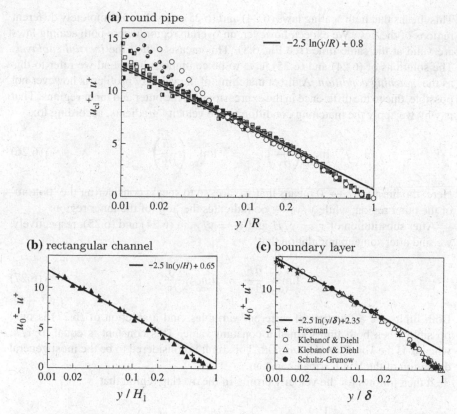

Fig. 6.9 The defect law for **a** turbulent pipe flow (with $R = \frac{1}{2}D$), **b** turbulent flow in a rectangular channel (with dimensions $2H_1 \ll 2H_2$), and **c** turbulent boundary layer flow (with a boundary layer thickness δ). Symbols for the pipe flow data are identical to those in Fig. 6.3. Data in (**b**) and (**c**) are taken from Monin and Yaglom (1973)

of experimental data the following scaling law holds for the mean velocity profile in the inner region:

$$\frac{\overline{u}}{u_*} = f\left(\frac{y}{y_0}\right). \tag{6.25}$$

This equation is known as the '*law of the wall*'. It expresses how the velocity profile becomes *independent* of the flow geometry close to the wall, that is, it is independent of the length scale H. This is illustrated in Fig. 6.3 for a *smooth* wall, for which $y_0^+ = 0.135$, according to (6.20). We previously interpreted y_0 as being proportional to the thickness of the viscous sublayer.

For a turbulent wall flow it holds that $H/y_0 \gg 1$, because with the value for y_0 as mentioned above, it follows that

$$\frac{H}{y_0} \sim \frac{H u_*}{\nu} \gg 1.$$

This means that both scaling laws (6.24) and (6.25) are valid in completely different regions of the flow. We expect, however, an overlap region where both scaling laws are 'valid at the same time (see Fig. 6.8). This area is called the *inertial sublayer*. The solutions of (6.24) and (6.25) have to be continuous here, and we refer to this as the *matching condition*. A direct matching of the velocity profiles is however not possible, due to the difference in the expressions for the outer and inner regions. That is why we apply the matching condition to the velocity gradients, according to:

$$\lim_{y/H \to 0} \left(\frac{\partial \overline{u}}{\partial y} \right)_{outer} = \lim_{y/y_0 \to \infty} \left(\frac{\partial \overline{u}}{\partial y} \right)_{inner}. \tag{6.26}$$

Here, the limit $y/H \to 0$ means that we are, so to speak, considering the 'bottom' of the outer region, while $y/y_0 \to \infty$ indicates the 'top' of the inner region.

After substitution of $y_e = y/H$ and $y_i = y/y_0$ in (6.24) and (6.25), respectively, we find after some manipulation that

$$\lim_{y_e \to 0} y_e \frac{\partial F}{\partial y_e} = \lim_{y_i \to \infty} y_i \frac{\partial f}{\partial y_i}. \tag{6.27}$$

Both limits are a function of different variables, and a solution of (6.27) is only possible when both limits equal a constant value. This constant is equal to $1/k$, where k is the *Von Kármán constant*. This limit is considered to be the most general definition of the Von Kármán constant.

It then follows for the velocity profile in the overlap region that

$$F = \frac{1}{k} \ln y_e + A \Rightarrow \frac{\overline{u} - u_0}{u_*} = \frac{1}{k} \ln \left(\frac{y}{H} \right) + A, \tag{6.28}$$

$$f = \frac{1}{k} \ln y_i \Rightarrow \frac{\overline{u}}{u_*} = \frac{1}{k} \ln \left(\frac{y}{y_0} \right). \tag{6.29}$$

In the solution of f we take the integration constant equal to zero. We thus retrieve the logarithmic velocity profile in (6.22). However, the essence of this result is that the logarithmic velocity profile appears to be a direct consequence of the existence of two different scaling regions that partially overlap each other.

Equations (6.28–6.29) can be further elaborated by eliminating the velocity $\overline{u}$. It then follows that

$$\frac{u_0}{u_*} = \frac{1}{k} \ln \left(\frac{H}{y_0} \right) - A. \tag{6.30}$$

With $y_0 = 0.135\nu/u_*$ in (6.20), we can rewrite this equation as:

$$\frac{u_0}{u_*} = \frac{1}{k} \ln \left(\frac{H u_*}{\nu} \right) + 5.0 - A.$$

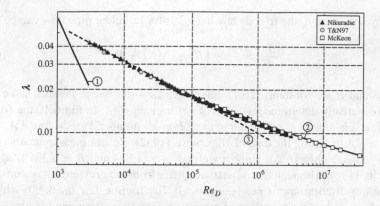

Fig. 6.10 Friction law for pipe flow, with the friction factor λ, defined in (6.34), as a function of the Reynolds number $Re_D = U_b D/\nu$, based on the bulk velocity U_b and pipe diameter D. The *lines represent*: *1* the friction factor for laminar Poiseuille flow ($\lambda = 64/Re_D$); *2* the friction law (for a smooth wall) in (6.35), with $A = 1.884$ and $B = 0.331$ (Zagarola and Smits 1998); *3* the Blasius friction law ($\lambda = 0.316 Re_D^{-1/4}$). Experimental data from: Nikuradse (1932), den Toonder and Nieuwstadt (1997), McKeon et al. (2004)

This expression is referred to as a *friction law*. Commonly, u_0/u_* relates to a *friction coefficient* c_f, defined as:

$$c_f \equiv \frac{\tau_s}{\frac{1}{2}\rho_0 u_0^2} = 2\left(\frac{u_*}{u_0}\right)^2, \tag{6.31}$$

where τ_s is again the wall shear stress, defined in (6.7).

The friction law for a turbulent pipe flow is illustrated in Fig. 6.10. For pipe flow it is common to use the average velocity U_b taken over the entire pipe cross section, given by

$$U_b = \frac{Q}{\frac{1}{4}\pi D^2}, \tag{6.32}$$

where Q is the *volume flow rate* (and $\dot{m} = \rho_0 Q$ is the *mass flow rate*). Hence, U_b is commonly referred to as the *bulk flow velocity*. The ratio of the bulk velocity and mean centerline velocity in a turbulent pipe flow is approximately given by

$$U_b/u_0 \cong 0.8167, \tag{6.33}$$

which is found by integrating a $\frac{1}{7}$th power velocity profile[1] (Schlichting 1979; White 2011). For a pipe flow with diameter D and length L, it holds that (White 2011)

$$-\frac{1}{\rho_0}\frac{d\overline{p}_0}{dx} = \frac{4u_*^2}{D} \quad \text{and} \quad \Delta\overline{p}_0 = \lambda \cdot \frac{L}{D} \cdot \frac{1}{2}\rho_0 U_b^2 \quad \Rightarrow \quad \lambda = 8\left(\frac{u_*}{U_b}\right)^2. \tag{6.34}$$

---

[1]For: $u(y) = u_0(y/R)^{1/n}$ we find for a pipe flow that: $U_b/u_0 = \frac{2n^2}{(2n+1)(n+1)}$.

Then, by substitution, the friction law in (6.30) for turbulent pipe flow can be written as:

$$\frac{1}{\sqrt{\lambda}} = A \log \left(Re_D \cdot \sqrt{\lambda} \right) - B, \tag{6.35}$$

where A and B are constants. This implicit expression is due to Prandtl, and can be used to iteratively determine the value of λ for given Re_D. In Fig. 6.10 the friction coefficient λ is plotted as a function of the Reynolds number. Prandtl found $A = 2.0$ and $B = 0.8$ by fitting the data of Nikuradse (1932). Recent measurements over a very large range in Reynolds number yielded: $A = 1.884$ and $B = 0.331$ (Zagarola and Smits 1998). Although these constants differ from the previous set of constants, the resulting friction law is nearly identical. The friction law in (6.35), which is essentially based on the equations of motion and the Prandtl mixing length closure, is valid over the full range of Reynolds numbers for which experimental data exist, i.e., between $Re_D = 3 \times 10^3$ and 36×10^6; see Fig. 6.10. An expression similar to (6.35) can be obtained for channel flow by rearranging (6.23). Using the constants: $k = 0.40$, $y_0^+ = 0.135$ ($\Pi = 2.0$), $\alpha = 0.16$, $\beta = 0.13$, and (6.33), one finds: $A = 1.67$ and $B = 0.5$.

Figure 6.10 also includes the *Blasius friction law* (Schlichting 1979):

$$\lambda = 0.316 \, Re_D^{-1/4}, \tag{6.36}$$

which is approximately valid for Reynolds numbers between 10^4 and 10^5. The Blasius friction law is frequently used, although it is based on the empirical $\frac{1}{7}$th power law for the mean velocity profile:

$$\frac{u(y)}{u_*} = C \left(\frac{y u_*}{\nu} \right)^{1/7} \Rightarrow \frac{u_0}{u_*} = C \left(\frac{R u_*}{\nu} \right)^{1/7} \Rightarrow \frac{U_b}{u_*} = C' \left(\frac{R u_*}{\nu} \right)^{1/7}$$

$$\Rightarrow \frac{U_b}{u_*} = C'' \left(\frac{R U_b}{\nu} \right)^{1/8},$$

with: $R = \frac{1}{2}D$, $U_b/u_0 = 0.8167$, and where C'' ($=1.217 \times 10^{-6}$) is an empirical constant (likewise C and C'). Thus, despite its attractively simple form, the Blasius friction law lacks a solid physical basis.

Problems

1. Consider a turbulent boundary layer. Suppose we are capable of doubling the thickness of the viscous sublayer. Will the friction increase or decrease?
2. Argue how neither y_0 nor D form a characteristic length scale in the overlap area between the outer and the inner region. The only scales that remain are u_* and y. Using dimensional analysis, derive a scaling law for the *velocity gradient* and show that this leads to a logarithmic velocity profile.

6.4 Wall Roughness

So far we did not consider the properties of the wall. Every wall and every wall material is characterized by (small) irregularities, which we refer to as *roughness*. These roughness elements have a characteristic height h (see Fig. 6.11). Examples of wall roughness include the sand and pebbles on the bottom of a riverbed ($h \sim$ 1–100 mm), landscapes covered by grass, crops, and the tree canopy of a forest ($h \sim$ 1 cm to 1–5 m). Apart from roughness with a random structure (Fig. 6.11a), one may also encounter more or less structured roughness, such as the inside of a drawn pipe (Fig. 6.11b) or a corrugated pipe, *riblets* on shark skin, a planned city ($h \sim$ 10–100 m), or sand dunes ($h \sim$ 10–500 m). Even surfaces that are considered to be 'smooth' have a finite roughness height, as illustrated in Fig. 6.11a, b.

Nikuradse made a thorough study of turbulent flows in pipes with rough surfaces by covering the inner walls of pipes with sand grains with a rather narrow size distribution, which provides a rather well-defined roughness scale. These measurements have been the basis of friction charts defined by Moody and Colebrook that can be found in almost any text book on fluid mechanics (e.g. White 2011).

Now consider the Reynolds number hu_*/ν, which can be interpreted as the ratio between the roughness height h and the thickness of the viscous sublayer, $\sim \nu/u_*$. When $hu_*/\nu < 1$, the roughness height is smaller than the thickness of the viscous sublayer. In that case the wall is considered to be *smooth*, and the mean velocity

(a) drawn aluminium pipe **(b)** steel pipe surface **(c)** shark skin

(d) forest canopy **(e)** cityscape **(f)** desert dunes

Fig. 6.11 Wall roughness of a characteristic height h ($\cong 4\,k_{rms}$). Examples of wall roughness: **a** surface (0.20×0.16 mm^2) of a drawn aluminum pipe: $k_{rms} = 0.16\,\mu$m (Shockling et al. 2006); **b** scanned surface (1.4×1.1 mm^2) of a non-rusted commercial steel pipe: $k_{rms} = 5\,\mu$m (Langelandsvik et al. 2008); **c** scales of the great white shark: $k_{rms} \cong 0.1$ mm (Bechert et al. 2000); Aerial views of **d** tropical forest in Gabon (photo J. Westerweel), and **e** Barcelona and **f** the Namib desert (*source* Google Earth)

profile in (6.21) remains valid. In other words, the roughness is too small to affect the mean velocity profile. The value of y_0 is given by (6.20), which was considered to scale with the thickness of the viscous sublayer.

The situation changes when $hu_*/\nu > 1$. The roughness now protrudes through the viscous sublayer, and the surface is now considered to be a *rough* wall. The flow in the vicinity of the wall is now mainly dominated by separating flow from the roughness elements and the generation of small detached eddies. This is essentially an inertial process, and therefore viscous effects are negligible. In other words, a viscous sublayer can no longer be identified.

Consequently, the integration constant y_0 is no longer proportional to ν/u_*, but to the roughness height h. On the basis of experimental data from pipe flows, where the wall was roughened using sand grains, it is found that:

$$y_0 = \frac{1}{30}h. \qquad (6.37)$$

For this reason, y_0 is sometimes referred to as the *roughness length*. In short, the integration constant y_0 is representative of the wall properties.

We now combine the velocity profiles for the smooth and the rough wall as follows:

$$\bar{u} = u_* \left\{ \frac{1}{k} \ln\left(\frac{y}{h}\right) + B' \right\}, \qquad (6.38)$$

where B' is a function of the Reynolds number hu_*/ν. The term B' is illustrated in Fig. 6.12. It has to meet the following asymptotic behavior:

$$B' = \begin{cases} 2.5 \ln\left(\dfrac{hu_*}{\nu}\right) + 5 & \text{for: } \dfrac{hu_*}{\nu} \ll 1 \text{ (i.e., a smooth wall),} \\[2ex] 8.5 & \text{for: } \dfrac{hu_*}{\nu} \gg 1 \text{ (i.e., a rough wall).} \end{cases} \qquad (6.39)$$

Fig. 6.12 The roughness parameter B' in (6.38) as a function of the dimensionless roughness hu_*/ν. After: Monin and Yaglom (1973)

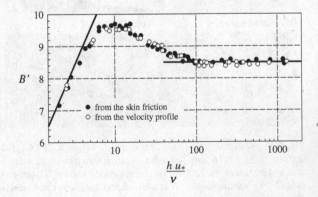

This last expression implies that the influence of viscosity on the wall region vanishes for a rough wall. Consequently, in the limit $hu_*/\nu \to \infty$ the thickness of the viscous sublayer becomes infinitely small. In that case every wall becomes eventually rough. Hence, given that in many technical applications and in nature the Reynolds numbers can be very high, smooth-wall turbulent flows are the exception, rather than the rule.

The asymptotes in (6.39) have been indicated by lines in Fig. 6.12. As a direct consequence of this expression for B' it follows that the friction law (6.30) becomes a function of hu_*/ν. The result of this being that all our scaling relations for the mean velocity profile, as discussed in Sect. 6.3, remain valid. The effect of roughness is, so to speak, absorbed in adjusting the constant in the logarithmic profile. In particular, the velocity profile in the outer region remains unchanged. This type of roughness is commonly referred to as *k-type roughness*. Other types of roughness do exist for which such simplifications does not hold. Take for example a grooved wall where the grooves appear at a regular spacing perpendicular to the direction of the flow. Then, under certain circumstances, the wall geometry does influence the velocity profile in the outer region; then the roughness is referred to as *d-type roughness*.

Although in general, a rough wall *increases* friction, or flow *drag*, it happens that special surfaces can effect a drag *reduction*. It appears that small *riblets* that are aligned with the flow direction can achieve a significant drag reduction up to 10%. The maximum drag reduction occurs for a rib spacing s between 14 and 20 viscous wall units, i.e., $14 < s^| < 20$, as illustrated in Fig. 6.13. This is an example of *bio-mimetics*, since the development and application of ribbed surfaces has been inspired by the presence of ribbed scales on sharks (Bechert et al. 2000); see Fig. 6.11c.

Problem

1. In Fig. 6.10 we find that the wall friction ρu_*^2 for a smooth pipe wall is a function of the Reynolds number. For a rough wall the wall friction eventually becomes independent of the Reynolds number. Explain this by generalizing (6.30) for a rough wall.

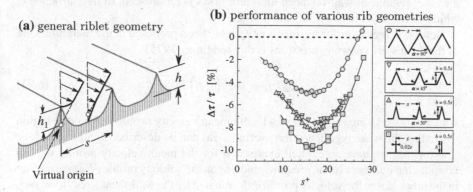

(a) general riblet geometry   **(b)** performance of various rib geometries

Fig. 6.13 **a** General geometry of a ribbed surface with a rib spacing s and rib height h. **b** Drag reduction performance of various rib geometries. From: Bechert et al. (2000)

6.5 Pressure Gradient

So far we have ignored the influence of any possible pressure gradients on the mean velocity profile. This implies that for the case of a turbulent boundary layer the results so far apply to a boundary layer over a flat plate with a constant-speed outer flow. In a channel flow or a pipe flow, however, a pressure gradient is evidently present, in order to force the fluid through the channel or pipe. It appears that, for a channel or pipe, the pressure gradient has no noticeable effect on the mean velocity profile, and thus its effect can be safely ignored. On the contrary, in boundary layer flows (in particular those over curved surfaces) pressure gradients can occur that cannot be neglected and that significantly affect the flow.

The question is, how does a pressure gradient modify the velocity profiles discussed in Sect. 6.3? Before this question can be answered, we first elaborate on the velocity profile in the outer region. Here, we limit ourselves to turbulent boundary layer flow.

The reference velocity u_0 is in this case the undisturbed velocity outside the boundary layer. For the length scale in the outer region we take the boundary layer thickness δ. In its general form the mean velocity profile in the outer region is given by the *defect law* (6.24). However, this is not really a suitable approach, since in practice we often need an explicit expression for the mean velocity profile.

When we consider the difference of the mean velocity profile in the outer region and the logarithmic velocity profile (6.29), it appears that this difference has a universal shape. This difference is referred to as the '*law of the wake*'. It is defined as

$$\frac{\overline{u}}{u_*} = \frac{1}{k} \ln\left(\frac{y}{y_0}\right) + \frac{2\Pi}{k} f_w\left(\frac{y}{\delta}\right), \qquad (6.40)$$

where Π is a constant, known as the *Coles parameter*. Incidentally, the expression '*law of the wake*' suggests that the function f_w resembles the profiles that are found in the wake flow behind bluff bodies. This underlines the notion that the turbulence in the outer regions of wall-bounded turbulent flows is reminiscent of *free turbulence*, which is discussed in Sects. 6.6 and 6.7.

Several expressions for the function f_w have been proposed in the literature. One of the most well-known expressions is due to Hinze (1975):

$$f_w(\eta) = \sin^2\left(\frac{\pi}{2}\eta\right), \qquad (6.41)$$

with $\eta = y/\delta$. Using (6.40) and (6.41), the mean velocity profile in the outer region (including the wake region and the overlap area) can be described completely.

Now that we have an explicit expression for the mean velocity profile, we can consider the influence of the pressure gradient on the velocity profile. We assume that in the inner region the velocity profile is dominated by the wall shear stress $\tau_s \equiv \rho u_*^2$. It then follows that in this area the mean velocity profile remains unaffected. Hence, the *law of the wall* (6.25) retains its universality.

The influence of the pressure gradient thus appears to be restricted to the outer region. The pressure gradient $d\overline{p}_0/dx$ forms, so to speak, an additional variable with which we can define a new dimensionless parameter. For this we take

$$\beta_* = \frac{\delta^*}{\tau_s}\frac{d\overline{p}_0}{dx}, \tag{6.42}$$

where δ^* represents the so-called *displacement thickness*, defined as (White 2011):

$$\delta^* = \int\limits_{0}^{\infty}\left(1 - \frac{\overline{u}}{u_0}\right)dy. \tag{6.43}$$

The parameter β_* is known as the *Clauser parameter*. When $\beta_* < 0$ the pressure *decreases* in the direction of the flow, which is referred to as a *favorable pressure gradient*. The opposite case, i.e., $\beta_* > 0$, is then referred to as an unfavorable, or *adverse*, pressure gradient; the fluid then flows against a rising pressure. Consequently, for increasing values of β_* the boundary layer eventually separates from the wall and reverses its direction (White 2011).

The *defect law* (6.24) in the outer region of the turbulent boundary layer is now generalized to:

$$\frac{\overline{u} - u_0}{u_*} = F\left(\frac{y}{\delta}, \beta_*\right). \tag{6.44}$$

It has to be noticed here that this profile is actually only valid for a so-called equilibrium boundary layer. This is a boundary layer for which β_* is constant. Only in that case it can be proven that the velocity profile retains its similarity as suggested by (6.44). However, in practice the profile described by (6.44) appears to provide a satisfactory description under other non-equilibrium circumstances as well.

There remains to exist an overlap region (Fig. 6.8) in which a logarithmic velocity profile (6.29) occurs. However, as β_* increases, this logarithmic region becomes increasingly thinner. In the limiting case of separating flow, when by definition the shear stress on the wall vanishes (i.e., $u_* = 0$), there is no logarithmic layer left. The velocity profile then appears to be proportional to y^2 (see Problem 1 below). The velocity profile for various values of the pressure gradient is illustrated in Fig. 6.14.

The attractiveness of the *law of the wake*, as it was described above, is that its dependence on β_* is fully described by the Coles parameter Π. On the basis of experimental data it is found that

$$\beta_* \approx -0.4 + 0.76\,\Pi + 0.42\,\Pi^2. \tag{6.45}$$

With this expression the mean velocity profile in a turbulent boundary layer exposed to a pressure gradient is completely described.

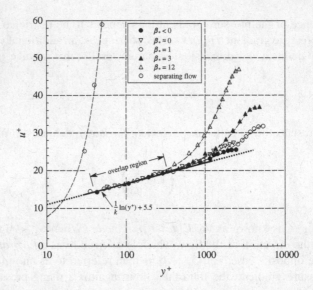

Fig. 6.14 The dimensionless mean velocity profile u_+ as a function of yu_*/ν for various values of the pressure gradient, expressed in terms of the *Clauser parameter*, β_*, defined in (6.42). The experimental data show the mean velocity profiles in a *favorable* pressure gradient ($\beta_* < 0$, $\Pi = 0.25$), a (nearly) *neutral* pressure gradient ($\beta_* \approx 0$, $\Pi = 0.6$), mild ($\beta_* \approx 1$, $\Pi = 1.1$), strong ($\beta_* \approx 3$, $\Pi = 2.2$), and very strong ($\beta_* \approx 12$, $\Pi = 4.5$) *adverse* pressure gradients, and a separating boundary layer. (Π is the *Coles parameter* in (6.40).) After: White (1991)

To finalize this section, we briefly discuss the friction coefficient c_f defined in (6.31). The effect of a pressure gradient on the value of c_f is commonly described by empirical correlations. The one that is most commonly used is due to Ludwieg and Tillmann (1950):

$$c_f = 0.246 \left(\frac{u_0 \theta}{\nu} \right)^{-0.268} 10^{-0.678H}, \tag{6.46}$$

where θ is the *momentum loss thickness*, and H the *shape factor*, defined as (White 2011)

$$\theta = \int_0^\infty \frac{\overline{u}}{u_0} \left(1 - \frac{\overline{u}}{u_0} \right) dy, \quad \text{and:} \quad H = \frac{\delta^*}{\theta}, \tag{6.47}$$

with δ^* given in (6.43).

Problems

1. Consider a boundary layer over a smooth flat plate exposed to a pressure gradient $d\overline{p}_0/dx > 0$. The wall shear stress equals $\tau_s \equiv \rho_0 u_*^2$ and the free-stream velocity outside the boundary layer is equal to u_0.

(a) Define a length scale L_p as

$$L_p^{-1} = \frac{1}{\tau_s} \frac{d\overline{p}_0}{dx}.$$

Discuss the physical background of this length scale. (Notice the resemblance to the *Obukhov length* defined in (B.3).)

(b) Discuss the velocity profile in the boundary layer as a function of y/L_p.

(c) Derive an explicit expression for the mean velocity profile for $y/L_p \gg 1$.

(d) Determine the shape of the velocity profile for $\tau_s = 0$.

2. Calculate the value of β_* in a channel flow and interpret the result.

6.6 Free Turbulent Flows

In this section we focus on the other type of turbulent flow: *free turbulent flows*. At the beginning of this chapter this was defined as unbounded turbulent flow, i.e., turbulence that is *not* influenced by the presence of a solid wall. The most important examples are the *free jet*, the *wake flow* behind a bluff body, and the turbulent *mixing layer*. These examples are illustrated in Fig. 6.15.

In all examples of Fig. 6.15 the flow develops as a function of the x-coordinate. However, this development is slow. In other words, the length scale of the flow field in the x-direction is much larger than the length scale in the y-direction. This implies that the gradients in the x-direction are much smaller than those in the y-direction, i.e.,

$$\frac{\partial}{\partial x} \approx \frac{1}{L} \ll \frac{\partial}{\partial y} \approx \frac{1}{\ell}, \tag{6.48}$$

or $\ell \ll L$. This geometrical condition enables us to reduce the Reynolds-averaged continuity equation (5.18) and Navier-Stokes equations (5.20) to a set of equations that is referred to as the *boundary layer equations*. For the case of a two-dimensional flow (for example, a *planar* jet, the wake behind a very long cylinder, or the mixing layer behind a flat splitter plate), the complete equations of motion for a stationary average flow then read:

$$\frac{\partial \overline{u}}{\partial x} + \frac{\partial \overline{v}}{\partial y} = 0, \tag{6.49}$$

$$\overline{u} \frac{\partial \overline{u}}{\partial x} + \overline{v} \frac{\partial \overline{u}}{\partial y} = -\frac{1}{\rho_0} \frac{\partial \overline{p}}{\partial x} - \frac{\partial}{\partial x}\left(\overline{u'^2}\right) - \frac{\partial}{\partial y}\left(\overline{u'v'}\right), \tag{6.50}$$

$$\overline{u} \frac{\partial \overline{v}}{\partial x} + \overline{v} \frac{\partial \overline{v}}{\partial y} = -\frac{1}{\rho_0} \frac{\partial \overline{p}}{\partial y} - \frac{\partial}{\partial y}\left(\overline{v'^2}\right) - \frac{\partial}{\partial x}\left(\overline{u'v'}\right). \tag{6.51}$$

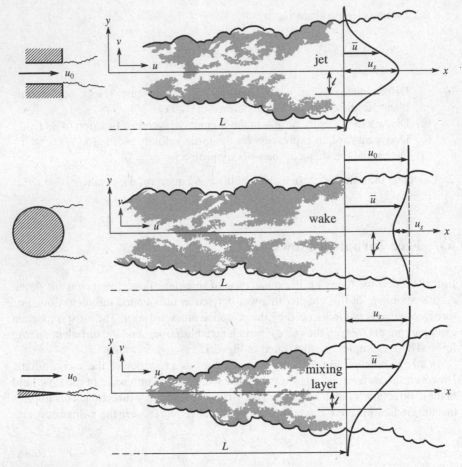

Fig. 6.15 Examples of free turbulent flows: the turbulent jet, wake, and mixing layer, including the definitions of the coordinate systems and of the length scales and velocity scales. After: Tennekes and Lumley (1972)

In these equations the molecular terms have also been ignored (see the discussion in Sect. 5.3). We also disregard any density effects.

In order to estimate the magnitude (and thus importance) of the various terms in (6.49–6.51), we need to introduce an appropriate velocity scale. For the mean velocity $\overline{u}$ in the x-direction we define u_s (see Fig. 6.15) as the characteristic velocity scale. It is also assumed that the spatial variation in $\overline{u}$ scales with u_s. An example of such a condition is a jet flow, where often u_s is taken equal to the mean velocity at the jet axis (see Sect. 6.7). However, alternative conditions can occur in which the spatial variation in $\overline{u}$ is much smaller than $\overline{u}$ itself. An example of this is situation is found in Problem 1 below.

Using the continuity equation (6.49) and the expression (6.48) it follows that

$$\frac{\partial \overline{v}}{\partial y} \sim \mathcal{O}\left(\frac{v_s}{\ell}\right) = -\frac{\partial \overline{u}}{\partial x} \sim \mathcal{O}\left(\frac{u_s}{L}\right),$$

so that we find for the characteristic scale v_s of $\overline{v}$ that: $v_s = u_s \ell / L$. For the scaling of turbulence we use the macroscale $\mathcal{U}$. We now assume that the following relation exists between both velocity scales:

$$\frac{\mathcal{U}^2}{u_s^2} = \mathcal{O}\left(\frac{\ell}{L}\right). \tag{6.52}$$

This needs to be verified for every flow.

Given the characteristic scales for the means of the two velocity components, we now estimate the magnitude of the terms in Eq. (6.51) for the y-momentum. Under the condition (6.52) it follows that when the largest terms are retained:

$$0 = -\frac{1}{\rho_0}\frac{\partial \overline{p}}{\partial y} - \frac{\partial}{\partial y}\left(\overline{v'^2}\right).$$

Integration of this expression yields:

$$\overline{p} = p_0 - \rho_0 \overline{v'^2},$$

where p_0 represents the pressure outside the turbulent flow region. Here we assume that $p_0 \neq f(x)$, so that the pressure gradient term in (6.50) can be expressed as:

$$\frac{1}{\rho_0}\frac{\partial \overline{p}}{\partial x} = -\frac{\partial}{\partial x}\left(\overline{v'^2}\right).$$

Substitution in the equation of the x-momentum (6.50) leads to:

$$\overline{u}\frac{\partial \overline{u}}{\partial x} + \overline{v}\frac{\partial \overline{u}}{\partial y} = -\frac{\partial}{\partial x}\left(\overline{u'^2} - \overline{v'^2}\right) - \frac{\partial}{\partial y}\left(\overline{u'v'}\right).$$

Once more, we apply the scaling in (6.52) to this last equation, which indicates that the first term on the right is very small. (In addition, it is observed that $\overline{u'^2} \approx \overline{v'^2}$ in most free turbulent flows; see for example Fig. 6.19.) It now follows that

$$\overline{u}\frac{\partial \overline{u}}{\partial x} + \overline{v}\frac{\partial \overline{u}}{\partial y} = -\frac{\partial}{\partial y}\left(\overline{u'v'}\right). \tag{6.53}$$

This expression taken together with the continuity equation (6.49), describe a flow geometry for which the dimension in one direction is much smaller compared to the other directions. This has coined the name 'boundary layer equations'.

Fig. 6.16 Visualization of a turbulent mixing layer, taken by Brown and Roshko (1974), in simultaneous top view and side view. From: Van Dyke (1982)

For a three-dimensional axisymmetrical geometry, a similar set of equations can be found that describe the fluid motion in a cylindrical coordinate system with an axial coordinate x and a radial coordinate r. The result reads:

$$\frac{\partial \overline{u}}{\partial x} + \frac{1}{r}\frac{\partial}{\partial r}(r\overline{v}) = 0, \tag{6.54}$$

$$\overline{u}\frac{\partial \overline{u}}{\partial x} + \overline{v}\frac{\partial \overline{u}}{\partial r} = -\frac{1}{r}\frac{\partial}{\partial r}\left(r\overline{u'v'}\right). \tag{6.55}$$

For details on the derivation of these equations we refer to any of the standard text books from the literature list.

Given that the evolution of the flow in the x-direction is slow, the sizes of turbulent eddies do not directly depend on L. The dimensions of the eddies at position x are thus only determined by the local length scale $\ell(x)$ (see Fig. 6.15), which is for example proportional to the thickness of the turbulent region. This is further illustrated in Fig. 6.16.

Thus, ℓ appears to be the only characteristic length scale. This is an important difference from wall-bounded turbulence where different scaling regions are present with their own characteristic length scale. In particular, we can associate the length scale ℓ with the length scale H of the outer region, which we introduced in Sect. 6.3. That is why in many ways free turbulence is comparable with turbulence in the outer region. This was already noted in the discussion of the *law of the wake* in Sect. 6.5.

On the basis of the scaling arguments discussed above, we can write

$$\overline{u} = u_s\, f\left(\frac{y}{\ell}\right), \quad \text{and:} \quad -\overline{u'v'} = \mathcal{U}^2\, g\left(\frac{y}{\ell}\right). \tag{6.56}$$

Here, ℓ, $\mathcal{U}$ and u_s are essentially functions of x, expressing how turbulent eddies grow with increasing x (see Figs. 6.15 and 6.16). The principal difference with wall-bounded turbulent flows is that the expressions in (6.56) are valid everywhere in the flow. When we use a closure model that relates the Reynolds stress to the gradient of the mean velocity (see Sect. 5.4), the function g can be expressed in terms of the function f. This means that we can describe the structure of the flow, provided it is appropriately scaled, by a *single* function. This result is referred to as *self-similarity* or *self-preservation* of the flow.

The fact that the turbulent region increases in size for increasing x means that the external fluid, which is not yet turbulent (i.e., which is *irrotational*), has to be absorbed, so to speak, by the turbulent flow region. This process is called *entrainment*. This suggests that a clear distinction can be made between the turbulent flow region and the flow region that is not turbulent. This indeed appears to be the case when we look at the visualizations of the boundary layer in Fig. 6.7 and the mixing layer in Fig. 6.16. Here the turbulent and non-turbulent flow regions are clearly separated by a sharp boundary.

Another striking feature of Figs. 4.6 and 6.16 is that we can recognize in the mixing layer large-scale structures (associated with the macrostructure) in the presence of small-scale structures (i.e., the microstructure). In Sect. 4.2 we discussed the physical aspects of these different structures. In Fig. 6.16 the large-scale structure is clearly visible. The top view indicates that these large-scale vortices are two-dimensional. They remind us of the vortices that relate to the Kelvin–Helmholtz instability (Sect. 3.2), which is also responsible for the emergence of turbulence in this flow geometry. Such large-scale, regular and reproducible structures are observed in nearly all turbulent flows, and they are commonly referred to as '*coherent structures*'. A significant amount of research has been dedicated to these coherent structures, primarily aimed at a better understanding of their dynamics and thus at a better understanding of turbulence as a whole. We return to these structures Sect. 8.2.

Finally, it is noticed that the interface between the turbulent and non-turbulent flow regions is strongly convoluted with many larger and smaller bulges, as shown in Figs. 6.17 and 6.18. It has been suggested that this interface would have a *fractal* structure (Sreenivasan 1991), which we encountered before in Problem 4 of Sect. 4.2. A direct consequence of this particular shape is that a fixed measurement probe close to the interface would only measure a turbulent signal part of the time. This phenomenon is called *intermittency* and is thus attributable to the fact that a particular variable, in this case turbulence, is not uniformly distributed in space.

Problems

1. In deriving (6.53), we made use of the fact that $\bar{u} \sim u_s$. This may not always be the case. An example is the wake flow or a free jet in a fluid moving along with it, i.e., a so-called *coflowing jet*. In that case the velocity scale u_s that characterizes the average velocity in the turbulent region, is much smaller than the total average velocity. In other words, $\partial \bar{u}/\partial x \sim \mathcal{O}(u_s/L)$, but: $\bar{u} \gg u_s$. Show that in this case the equation for the x-momentum reduces to

Fig. 6.17 Visualization image of the far-field of a jet observed in a plane normal to the jet x-axis. The jet Reynolds number is 9×10^3. The fractal dimension of the turbulent/non-turbulent interface of a jet is approximately 2.4 (Sreenivasan 1991). Image from: Catrakis and Dimotakis (1996)

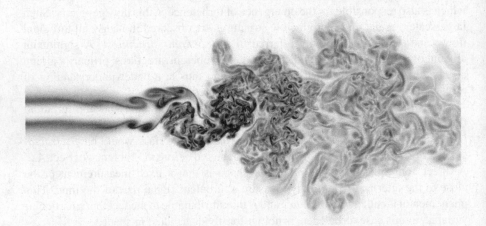

Fig. 6.18 Visualization of the vorticity magnitude of a round jet at $Re = 3 \times 10^3$ developing from laminar to turbulent flow. Notice the development of the shear layers in the laminar domain, akin the Kelvin–Helmholtz instability discussed in Sect. 3.2. Data are obtained from a direct numerical simulation (DNS) by Moore (2009)

$$U_0 \frac{\partial \overline{u}}{\partial x} = -\frac{\partial}{\partial y}\left(\overline{u'v'}\right),$$

where U_0 represents the mean flow speed outside the turbulent region.

2. Apply the scaling we used in this section to the equation

$$-\overline{u'v'} = K \frac{\partial \overline{u}}{\partial y},$$

with $K \sim \mathcal{UL}$.

   (a) Determine a scaling relation for $\mathcal{L}$.

   (b) Suppose that $L \sim x$. A typical value for ℓ/x is 6×10^{-2}. Use this to compute the magnitude of the other relevant scales.

6.7 The Free Jet

In this section we apply the boundary layer equations, derived in the previous section, to the problem of a turbulent free jet. We represent the jet in a three-dimensional axisymmetrical geometry. In that case the equations of motion are given by (6.54–6.55).

A submerged turbulent round jet is formed when a fluid is issued through a round nozzle with a diameter d into a stagnant environment of the same fluid with zero pressure gradient. It is assumed that the fluid in the nozzle has a uniform profile with a velocity u_0. We define a coordinate system with the x-axis in the direction of the flow and with its origin at the nozzle exit opening of the nozzle. The environment has an infinite size (i.e., there is no influence from any walls). The flow inside this domain develops into a free jet, as illustrated in Figs. 6.15 and 6.18.

The initial condition for this flow is determined by the momentum flux in the x-direction. The momentum flux is introduced at the position of the nozzle exit ($x = 0$), given by:

$$M_0 = \frac{\pi}{4}d^2\rho_0 u_0^2. \tag{6.57}$$

It is easily shown that the momentum is conserved in the x-direction, under the conditions stated in the previous paragraph (Rajaratnam 1976). For this we multiply (6.55) by $\rho_0 r$ and integrate with respect to r from $r = 0$ to ∞:

$$\int_0^\infty \rho_0 \overline{u}\, r \frac{\partial \overline{u}}{\partial x}\, dr + \int_0^\infty \rho_0 \overline{v}\, r \frac{\partial \overline{u}}{\partial r}\, dr = -\int_0^\infty \rho_0 \frac{\partial (r\,\overline{u'v'})}{\partial r}\, dr. \tag{6.58}$$

The three terms in the equation above can be expressed as

$$\int\limits_0^\infty \rho_0 \bar{u}\, r\, \frac{\partial \bar{u}}{\partial x}\, dr = \frac{1}{4\pi}\frac{d}{dx}\int\limits_0^\infty (\rho_0 \bar{u}^2)\, 2\pi r\, dr,$$

$$\int\limits_0^\infty \rho_0 \bar{v}\, r\, \frac{\partial \bar{u}}{\partial r}\, dr = \rho_0 \bar{u}\,\bar{v}\Big|_0^\infty - \int\limits_0^\infty \rho_0 \bar{u}\,\frac{\partial (r\,\bar{v})}{\partial r}\, dr = \int\limits_0^\infty \rho_0 \bar{u}\,\frac{\partial (r\,\bar{u})}{\partial x}\, dr$$

$$= \frac{1}{4\pi}\frac{d}{dx}\int\limits_0^\infty (\rho_0 \bar{u}^2)\, 2\pi r\, dr,$$

$$\int\limits_0^\infty \rho_0 \frac{\partial (r\,\overline{u'v'})}{\partial r}\, dr = \rho_0 r\,\overline{u'v'}\Big|_0^\infty = 0.$$

For the second term we applied the continuity equation (6.54), and we applied the boundary conditions:

$$\bar{u} = 0 \text{ for: } r \to \infty, \quad \bar{v} = 0 \text{ for: } r = 0, \quad \overline{u'v'} = 0 \text{ for: } r = 0 \text{ and } r \to \infty.$$

Then, (6.58) becomes:

$$\frac{d}{dx}\int\limits_0^\infty (\rho_0 \bar{u}^2)\, 2\pi r\, dr = 0, \tag{6.59}$$

which states that the x-momentum is *conserved*. We call this condition, which must be satisfied by the flow for every x-position, an *integral property* of the flow.

Because the mean flow is axisymmetric we can introduce a *stream function* ψ, defined as:

$$\bar{u} = \frac{1}{r}\frac{\partial \psi}{\partial r}, \quad \bar{v} = -\frac{1}{r}\frac{\partial \psi}{\partial x}. \tag{6.60}$$

With this, the continuity equation (6.54) is implicitly satisfied. Next, we apply the *similarity condition* (6.56). It follows that

$$\psi = u_s\, \ell^2 F\left(\frac{r}{\ell}\right), \quad \text{and: } -\overline{u'v'} = \mathcal{U}^2 g\left(\frac{r}{\ell}\right), \tag{6.61}$$

where the scales $u_s, \mathcal{U}$ and ℓ are functions of x only. With (6.60) it now follows that

$$\bar{u} = u_s\, \frac{F'(\eta)}{\eta} = u_s f(\eta), \tag{6.62}$$

with: $\eta = r/\ell(x)$. We normalize the function $f(\eta)$ using $f(0) = 1$. This implies that u_s represents the velocity at the jet axis. Substitution in (6.59), and integration with the given initial condition (6.57), leads to the integral equation for the momentum M:

$$M \equiv \int_0^\infty \rho_0 \overline{u}^2 \, 2\pi r \, dr = 2\pi \, \rho_0 \, u_s^2 \, \ell^2 \int_0^\infty \eta \, f^2(\eta) \, d\eta = M_0. \qquad (6.63)$$

The integral of ηf^2 is equal to a constant, so that we find that the product $u_s \, \ell$ is *constant*. This implies that the Reynolds number of a round jet has a *constant* value for *all* x, and that it is essentially determined by the Reynolds number of the flow at the nozzle exit. Application of (6.60) and (6.61) now leads to

$$\overline{v} = u_s \frac{d\ell}{dx} \left\{ F' - \frac{F}{\eta} \right\} \qquad (6.64)$$

$$\frac{\partial \overline{u}}{\partial x} = \frac{du_s}{dx} \frac{F'}{\eta} - \frac{u_s}{\ell} \frac{d\ell}{dx} \eta \frac{d}{d\eta} \left(\frac{F'}{\eta} \right) \qquad (6.65)$$

$$\frac{\partial \overline{u}}{\partial r} = \frac{u_s}{\ell} \frac{d}{d\eta} \left(\frac{F'}{\eta} \right). \qquad (6.66)$$

Substitution of these expressions in the equation of motion for $\overline{u}$, and after applying:

$$\frac{1}{u_s} \frac{du_s}{dx} = -\frac{1}{\ell} \frac{d\ell}{dx},$$

which follows from $u_s \ell = constant$, we arrive at:

$$-\frac{u_s^2}{\mathcal{U}^2} \frac{d\ell}{dx} \left\{ \left(\frac{F'}{\eta} \right)^2 + \frac{F}{\eta} \frac{d}{d\eta} \left(\frac{F'}{\eta} \right) \right\} = \frac{1}{\eta} \frac{d}{d\eta} (\eta g), \qquad (6.67)$$

where we used (6.61) for $-\overline{u'v'}$. A similarity solution is *only* possible when:

$$\frac{u_s^2}{\mathcal{U}^2} \frac{d\ell}{dx} = c = \text{constant}. \qquad (6.68)$$

This matches the assumption in (6.52) of which we mentioned that it needed to be verified.

There are now two possibilities to move forward. The first, and most common one, is that an assumption is made on the ratio $u_s/\mathcal{U}$. The term u_s is the mean velocity at the jet axis, while $\mathcal{U}$ is a characteristic velocity scale of the turbulence. For example, take $\mathcal{U}^2 = \frac{2}{3}e_0$, where e_0 represents the turbulent kinetic energy at the jet axis, with e defined as (see Chap. 7):

$$e = \frac{1}{2} \overline{u_i' u_i'}.$$

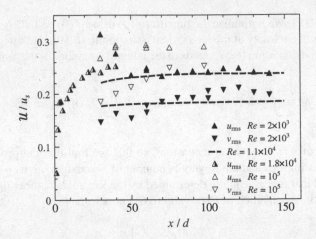

Fig. 6.19 The turbulence intensities $u_{\mathrm{rms}} = \sqrt{\overline{u'^2}}$ and $v_{\mathrm{rms}} = \sqrt{\overline{v'^2}}$ as functions of x/d at the axis of a free jet. Experimental data from: Fukushima et al. (2002) at $Re = 2 \times 10^3$ (*closed symbols*), Panchapakesan and Lumley (1993) (*broken lines*) at $Re = 1.1 \times 10^4$, Tong and Warhaft (1995) at $Re = 1.8 \times 10^4$, and Wygnanski and Fiedler (1969) at $Re = 10^5$ (*open symbols*)

In Fig. 6.19 are shown experimental data for the components of e as a function of x/d for three jets at different Reynolds numbers. It follows that e_0/u_s^2, and thus also $u_s/\mathcal{U}$, approaches a constant value for $x/d > 60$. It is often assumed that this constant is universal, that is, the ratio $u_s/\mathcal{U}$ does not depend on the details of the flow, and is thus would be the same for *all* jet flows. In other words, u_s is directly proportional to $\mathcal{U}$, and we only need to consider a *single* velocity scale in our problem, for which we use u_s. (Note, however, that the experimental data in Fig. 6.19 indicate a (weak) dependence of $u_s/\mathcal{U}$ on Reynolds number; we return to this aspect at the end of this section.)

For $\mathcal{U}/u_s = constant$, (6.68) gives:

$$\ell = \left(\frac{\mathcal{U}}{u_s}\right)^2 (x - x_0), \tag{6.69}$$

where x_0 represents the so-called *virtual origin* for which the similarity solution is valid (from experimental data it follows that $x_0/d \approx 0.5\text{--}10$, depending on the flow condition at the nozzle). Also, the constant c in (6.68) has been absorbed into the ratio $u_s/\mathcal{U}$ without loss of generality. Consequently, we notice that $\ell(x)$ must be universal as well. In other words, the jet spreading rate is identical for *all* jet flows. This can be verified by comparing the jets pictured in Fig. 6.23.

We can now integrate (6.67) to find

$$\eta g + \frac{F F'}{\eta} = A. \tag{6.70}$$

The integration constant A equals zero, given the boundary conditions that both $g = 0$ and $F'/\eta = 0$ for $\eta \to \infty$. We now reached the point at which we cannot continue without a closure hypothesis. On the basis of K-theory, we find

$$-\overline{u'v'} \equiv \mathcal{U}^2 \, g(\eta) = K \frac{\partial \overline{u}}{\partial r} = K \frac{u_s}{\ell} \frac{d}{d\eta} \left(\frac{F'}{\eta} \right). \qquad (6.71)$$

The self-similarity hypothesis then implies that:

$$\frac{K u_s}{\mathcal{U}^2 \ell} = B = \text{constant}. \qquad (6.72)$$

Substitution in (6.70) leads again to an integrable equation with the solution:

$$-\frac{1}{2} F^2 = B \left(\eta \, F' - 2F \right) + C. \qquad (6.73)$$

Here C is an integration constant that equals zero, since $F = 0$ for $\eta = 0$. (This boundary condition follows from the fact that the jet axis is also a streamline.) Further integration of (6.73) yields:

$$F = \frac{2D\eta^2}{1 + D\eta^2/(2B)},$$

from which it follows that

$$f(\eta) = \frac{F'}{\eta} = \frac{4D}{\left[1 + D\eta^2/(2B)\right]^2}.$$

Here D is again an integration constant, which can be evaluated using the normalization condition $f(0) = 1$ (i.e., $D = \frac{1}{4}$). This leads to the final solution for the mean axial velocity profile of a free self-similar jet

$$\frac{\overline{u}}{u_s} = \left(1 + \frac{\eta^2}{8B} \right)^{-2}, \qquad (6.74)$$

with: $\eta = r/\ell(x)$. The constant B can be determined from experimental data. First we define ℓ as $r_{1/2}$, that is, ℓ equals the radial distance where $\overline{u} = \frac{1}{2} u_s$ (or, *jet half-width*). From the experimental data in Fig. 6.20 we obtain:

$$\ell = 0.0965 \, (x - x_0). \qquad (6.75)$$

On the basis of (6.69) this means that: $\mathcal{U}^2/u_s^2 \cong 0.0965$. This is consistent with the data presented in Fig. 6.19, given the earlier definition $\mathcal{U} = (\frac{2}{3} e_0)^{1/2}$. (For the jet

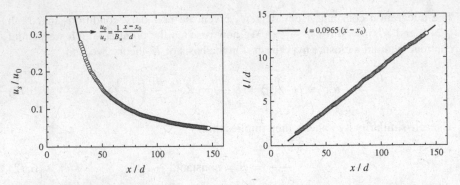

Fig. 6.20 The mean centerline velocity u_s relative to the nozzle exit velocity U_0 and the jet half-width ℓ relative to the nozzle diameter d of a round jet as a function of the distance x from the nozzle exit. The jet Reynolds number is 2×10^3. Experimental data from: Fukushima et al. (2002)

centerline we take $\overline{v'^2} = \overline{w'^2}$, so that: $e_0 = \frac{1}{2}\overline{u'^2} + \overline{v'^2}$.) Application of the definition $\ell = r_{1/2}$ in the solution (6.74) then gives:

$$B = 0.302.$$

Together with (6.71) this gives for the eddy viscosity K:

$$K \approx 0.302\, u_s \left(\frac{\mathcal{U}}{u_s}\right)^4 (x - x_0) = 0.00281\, u_s (x - x_0). \tag{6.76}$$

The solution with this value for K is plotted in Fig. 6.21. We see that this solution (in particular for small values of $r/(x - x_0)$) is in very good agreement with experimental data. For larger values of $r/(x - x_0)$ the theoretical solution appears to overestimate the velocity in comparison with the experimental data. This could be explained by the occurrence of *intermittency* that was discussed in the previous section. Because of this, the effective eddy viscosity K is reduced in the outer region of the jet. A correction of K for intermittency indeed leads to a better agreement of the theoretical result with experimental data. Note that the present analysis yields an eddy viscosity K that is valid over the *entire* flow domain; this then implies that the *external* irrotational (viz., non-turbulent) fluid has a finite non-zero value for K, which is strictly speaking physically incorrect.

Finally we compute a representative length scale on the basis of the solution above, based on: $K \sim \mathcal{U}\mathcal{L}$. We use (6.69), (6.76), and $\mathcal{U}/u_s \approx 0.311$, to find that:

$$\mathcal{L} = 0.094\,\ell.$$

In other words, $\mathcal{L} \ll \ell$. In light of our discussion in Sect. 5.4 this result explains the reasonable success of K-theory for the description of a free jet. This is illustrated in Fig. 6.22; compare a similar picture of the mixing layer in Fig. 6.16, where $\mathcal{L} \approx \ell$.

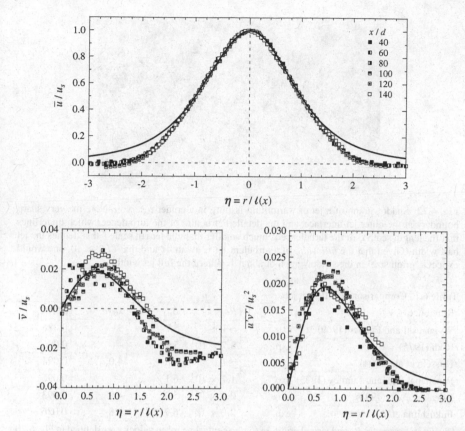

Fig. 6.21 The normalized profiles of the mean axial velocity $\overline{u}$, the mean radial velocity $\overline{v}$, and Reynolds stress $\overline{u'v'}$ of a self-similar free jet. The solid lines represent the solutions in (6.74), (6.64), and (6.71), respectively, with ℓ given by (6.75) and K given by (6.76). The symbols represent experimental data from Fukushima et al. (2002) of a $Re = 2 \times 10^3$ jet

We stated previously that there exists *two* possibilities to solve the equations of motion for a free jet on the basis of Eq. (6.68). In the alternative approach we assume that $d\ell/dx$ is *constant*, although it can still depend on the details of the flow, such as the initial conditions (George 1989). This is for example the shape of the velocity profile in the nozzle exit, which can either be laminar or turbulent. This means that $d\ell/dx$ can vary from experiment to experiment, and in this case the spreading rate of a free jet is *not* universal. The ratio between u_s and $\mathcal{U}$ would then follow from the application of (6.68). The subsequent analysis would be along a similar path as described above. An indication for the dependence on initial conditions at the jet nozzle are the different values for $\mathcal{U}/u_s$ in Fig. 6.19, as one jet develops from a laminar Poiseuille flow at the nozzle, while the other jet develops from a laminar uniform profile. Table 6.1 summarizes the decay rates B_u of the mean centerline velocity and virtual origins x_0 of different jet experiments and simulations. Please take note that variation of B_u and $d\ell/dx$ may also result as a consequence of realistic boundary

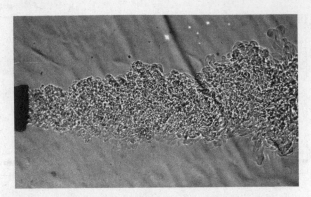

Fig. 6.22 Shadowgraph of a jet of warm fluid issuing in a colder reservoir. Note the very sharp boundary at the outer jet interface, and the detailed structure of the turbulence, which underlines that the length scale $\mathcal{L}$ of the turbulence is much smaller than the length scale ℓ that defines the jet half width (Given that the *full width* of a turbulent jet is about 4ℓ and that $\ell/\mathcal{L} \approx 10$, one would expect a 'grain size' in the shadowgraph of about 1/40th of the full jet width)

Table 6.1 Comparison of different jets

Reference		Re	B_u	x_0/d	$d\ell/dx$
Wygnanski and Fiedler (1969)	exp.	$\sim 10^5$	5.7	3	0.086
Rodi (1975)	exp.	8.7×10^4	5.9	–	0.086
Hussein et al. (1994)	exp.	$\sim 10^5$	5.8	4.0	0.094
Panchapakesan and Lumley (1993)	exp.	1.1×10^4	6.1	–	0.096
Boersma et al. (1998)	sim.	2.4×10^3	5.9	4.9	0.093
Fukushima et al. (2002)	exp.	2.0×10^3	6.7	6.8	0.096

The decay parameter B_u and virtual origin x_0 for the centerline mean velocity are defined in Fig. 6.20 (see also Problem 1 below): exp = experimental; sim = simulation. Adopted from: Boersma et al. (1998)

conditions imposed by the presence of wall at a finite distance. The presence of walls reduces the jet growth rate and induces a back flow of the external fluid, which both affect the results for B_u and $d\ell/dx$. In general the effects of walls can be ignored when the cross section of external domain is at least 10^5 times the cross section $\frac{\pi}{4}d^2$ of the nozzle.

Problems

1. For the velocity at the axis of the free jet the following equation can be derived:

$$u_s = U_0 \frac{B_u d}{x - x_0}$$

(see also Fig. 6.20). Compute the theoretical value for the constant B_u. Compare this result with the values reported in Table 6.1.

2. A free jet in two dimensions, or *planar jet*, is described by the following equations of motion:

$$\frac{\partial \overline{u}}{\partial x} + \frac{\partial \overline{v}}{\partial y} = 0$$

$$\overline{u}\frac{\partial \overline{u}}{\partial x} + \overline{v}\frac{\partial u}{\partial y} = -\frac{\partial}{\partial y}\overline{u'v'}.$$

Find the following similarity relations for this flow:

$$\ell \sim (x - x_0), \quad u_s \sim (x - x_0)^{-1/2}, \quad \text{and:} \quad \overline{u} = u_s\frac{1}{\cosh^2(\eta/\sqrt{2})},$$

with: $\eta = y/\ell(x)$.

3. In Problem 1 of Sect. 6.6 it was found that the equation for two-dimensional wake flow reads

$$U_0\frac{\partial \overline{u}}{\partial x} = -\frac{\partial \overline{u'v'}}{\partial y}.$$

Find the following similarity relations for this flow:

$$\ell \sim (x - x_0)^{1/2}, \quad u_s \sim (x - x_0)^{-1/2}, \quad \text{and:} \quad \overline{u} = u_s\, e^{-\frac{1}{2}B\eta^2},$$

where B is a constant, and $\eta = y/\ell(x)$. Solve the same problem for the three-dimensional wake.

4. A free jet flows with an initial speed U_0 along the x-axis in a fluid moving along with a velocity of U_1 (i.e., a co-flowing jet). It is given that: $U_0 \gg U_1$.

   (a) Discuss the development of this free jet at the initial and final phase in a qualitative manner.
   (b) Define the velocity u^* as:

   $$\overline{u} = U_1 + u^*.$$

   Derive on the basis of the continuity equation that the total mass m of the jet fluid increases at a rate:

   $$\frac{\partial m}{\partial x} \equiv \frac{\partial}{\partial x}\int_0^\infty \rho_0(U_1 + u^*)\, 2\pi r\, dr = E,$$

   where E, which is called the *entrainment rate*, is defined as

   $$E = -2\pi\rho_0[r\,\overline{v}]_\infty.$$

(c) Show from the momentum equation that for the total momentum M of the jet:

$$\frac{\partial M}{\partial x} \equiv \frac{\partial}{\partial x} \int\limits_0^\infty \rho_0(U_1 + u^*)u^* \, 2\pi r \, dr = 0.$$

(d) Use the following similarity relation for u^*:

$$u^* = u_s \, f\left(\frac{r}{\ell}\right),$$

with: $f \equiv 1$ for $r \leqslant \ell$, and: $f = 0$ for $r > \ell$ (i.e., the so-called *top-hat profile*). In connection with this we take the following hypothesis for the entrainment rate:

$$E = \beta \cdot 2\pi \, \rho_0 \, \ell \, u_s$$

(a) $Re = 2 \times 10^3$ **(b)** $Re = 2 \times 10^8$

Fig. 6.23 **a** Visualization of a submerged round free jet with a Reynolds number of 2×10^3. The jet is issued from 1 mm diameter nozzle, and the displayed image height corresponds to 45 mm. Image taken by: Fukushima et al. (2002). **b** Photograph of the exhaust plume of a TITAN IV rocket discharging upward during a ground-based test. The bright plume is approximately 120 m high atop a 30 m test stand. The Reynolds number is about 2×10^8. Details are given by Mungal and Hollingsworth (1989). Use a compass or protractor to compare the jet spreading rates

where β is a constant. Provide an interpretation of this expression for the entrainment.

(e) Formulate two ordinary differential equations for ℓ and u_s. Show that for small values of x, when $u_s \gg U_1$, the solutions of these equations are consistent with the results for a free jet issuing into stagnant fluid. What is the solution for large values of x when $u_s \ll U_1$?

(f) Determine the complete solution for the differential equations of ℓ en u_s.

5. The hot exhaust gasses of the jet shown in Fig. 6.23b evidently have a much lower density than the surrounding air. The *Morton length* ℓ_M is the distance from the jet nozzle over which the buoyancy effects are negligible, and is defined by:

$$\ell_M = \frac{M_0^{3/4}}{B_0^{1/2}}, \quad \text{with:} \quad M_0 = Q_0 u_0, \quad \text{and:} \quad B_0 = Q_0 g \frac{\Delta \rho}{\rho},$$

where Q_0 is the volume flow rate at the nozzle. Estimate the Morton length for the jet in Fig. 6.23. The TITAN IV rocket engine delivers a thrust of 7×10^6 N at a mass flow rate of 268×10^3 kg in 2 min.

Chapter 7
Kinetic Energy

In Chap. 4 we stated that turbulence is dissipative, in other words, turbulence converts kinetic energy into heat. In this chapter we reinforce this statement using the equations for kinetic energy. In this way we will also gain a better understanding of the dynamics of turbulence.

7.1 Kinetic Energy of the Mean Flow

Before focusing on the kinetic energy of turbulence, we first consider the mean flow. We define the kinetic energy per unit mass as $E = \frac{1}{2}\bar{u}_i^2$ (pay attention to the summation convention as introduced in Sect. 2.1). An equation for E is found by multiplying Eq. (5.20) for the mean velocity $\bar{u}_i$ with $\bar{u}_i$. When we neglect density and temperature effects, it follows that:

$$\frac{DE}{Dt} \equiv \frac{\partial E}{\partial t} + \bar{u}_j \frac{\partial E}{\partial x_j} = P_u + T_u + D_u, \tag{7.1}$$

with:

$$P_u = -\frac{1}{\rho_0}\bar{u}_i \frac{\partial \bar{p}}{\partial x_i},$$

$$T_u = \frac{\partial}{\partial x_j}\left(-\bar{u}_i \overline{u'_i u'_j}\right), \tag{7.2}$$

$$D_u = \overline{u'_i u'_j} \frac{\partial \bar{u}_i}{\partial x_j}.$$

The left-hand side of (7.1) describes how the kinetic energy E changes for a point that moves at an average speed of $\bar{u}_i$. The right-hand side consists of three parts, which we discuss below.

© Springer International Publishing Switzerland 2016
F.T.M. Nieuwstadt et al., *Turbulence*, DOI 10.1007/978-3-319-31599-7_7

P_u: **production** This first term represents the work that is done by the mean pressure gradient to maintain the mean flow. (In some flow configurations the flow is driven by the boundary condition, e.g. a moving wall.) It is therefore obvious to call this term the production term of mean kinetic energy. In the case of a channel flow, we saw in Sect. 6.1 that $d\overline{p}/dx < 0$. An equilibrium must exist between the pressure gradient and the wall shear stress that relates to $\overline{u} > 0$ in the channel; in other words: $P_u > 0$.

T_u: **transport** This term has the form of a *divergence*. In the remainder of this book we will encounter divergence terms in other equations as well, which is why we discuss here the effect of this term in more detail. We can rewrite the divergence term as:

$$T_u = \frac{\partial F_j}{\partial x_j}, \quad \text{with:} \quad F_j = -\overline{u}_i \, \overline{u_i' u_j'}. \tag{7.3}$$

We can interpret F_j as the flux. Using the *divergence theorem*, it follows that

$$\iiint\limits_V \frac{\partial F_i}{\partial x_i} \, dV = \iint\limits_S F_j n_j \, dS, \tag{7.4}$$

where V is a volume of the flow enclosed by the surface S, with n_j as the outside normal to the surface. We now choose V so that at the boundary S the flux $F_i = 0$. This can be done in various ways: for example, we can choose S at a fixed wall, where it often holds that $F_i = 0$; or we choose S in an area where the mean flow or the turbulence equals zero. The result is that the right-hand side of (7.4) equals zero. This can be interpreted as follows. The transport term (7.3) can only result in a redistribution inside the volume V without a net increase or decrease. For this reason, (7.3), and thus T_u, is sometimes called the *redistribution term* or *transport term*.

D_u: **deformation work** This term can be understood with help of Eq. (5.26), with which we have coupled the Reynolds stress to the average strain rate tensor according to the molecular analogy. If we also use the fact that $\overline{u_i' u_j'}$ is symmetric, it follows that:

$$D_u \equiv \overline{u_i' u_j'} \frac{\partial \overline{u}_i}{\partial x_j} = \frac{1}{2} \overline{u_i' u_j'} \left(\frac{\partial \overline{u}_i}{\partial x_j} + \frac{\partial \overline{u}_j}{\partial x_i} \right)$$

$$= -\frac{K}{2} \left(\frac{\partial \overline{u}_i}{\partial x_j} + \frac{\partial \overline{u}_j}{\partial x_i} \right)^2 \leqslant 0. \tag{7.5}$$

This term is thus always negative; in other words, this term is a sink for the kinetic energy in the flow. The mechanism is, in this case, the *deformation work* that is performed by the *Reynolds stress*. Indeed,

$$\frac{1}{2} \left(\frac{\partial \overline{u}_i}{\partial x_j} + \frac{\partial \overline{u}_j}{\partial x_i} \right)$$

is the deformation rate tensor of the mean flow, and (7.5) can be interpreted as the work that has to be done by the Reynolds stress to deform the fluid elements.

The viscous terms that occur in the equation for E are negligible on the basis of the usual scaling using the Reynolds number, which is based on the macrostructure.

To illustrate the use of the energy equation, we apply the equation to the plane channel flow discussed in Chap. 6. The flow is in this case stationary and horizontally homogeneous. Equation (7.1) then reduces to:

$$0 = -\frac{1}{\rho_0}\overline{u}\frac{\partial \overline{p}}{\partial x} + \frac{\partial}{\partial y}\left(-\overline{u}\,\overline{u'v'}\right) + \overline{u'v'}\frac{\partial \overline{u}}{\partial y}. \tag{7.6}$$

In Fig. 7.1 the various terms in this equation are shown schematically as a function of the distance to the wall. For this, we used the profile for $\overline{u}$ and for $-\overline{u'v'}$ as derived in the previous chapter.

We observe that E is mostly produced in the middle of the channel, while the deformation work is done primarily close to the wall. Indeed, in the inner region, (6.15) and $-\overline{u'v'} \approx u_*^2$ hold, so that the following holds for the deformation work:

$$\overline{u'v'}\frac{\partial \overline{u}}{\partial y} \approx -\frac{u_*^3}{ky}.$$

We find that the deformation work strongly increases when approaching the wall. The difference between the production (in the center) and the loss (at the wall) is compensated by the transport term, which provides the exchange of energy between both regions in the flow.

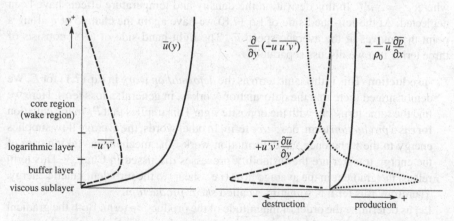

Fig. 7.1 Terms in Eq. (7.1) for E for channel flow (*Note* the profile of $\overline{u'v'}$ has been exaggerated with respect to $\overline{u}(y)$.)

7.2   Kinetic Energy of Turbulence

The kinetic energy (per unit mass) of the turbulent velocity fluctuations is defined as $e = \frac{1}{2}\overline{u_i'^2}$, thus again summed over the three coordinate directions. When the kinetic energy is scaled with the mean velocity $\overline{u}$, it is referred to as the *turbulence intensity*:

$$i = \frac{e}{\overline{u}^2}. \tag{7.7}$$

An equation for e is found by multiplying Eq. (5.21) for u_i' with u_i' and then by applying the standard Reynolds averaging procedure. The result is:

$$\frac{De}{Dt} \equiv \frac{\partial e}{\partial t} + \overline{u}_j \frac{\partial e}{\partial x_j} = P_k + T_k + \Pi_k + D_k - \epsilon, \tag{7.8}$$

with:

$$P_k = -\overline{u_i' u_j'} \frac{\partial \overline{u}_i}{\partial x_j},$$

$$T_k + \Pi_k + D_k = \frac{\partial}{\partial x_j} \left(-\overline{u_j' e'} - \frac{1}{\rho_0} \overline{p' u_j'} + \nu \frac{\partial e}{\partial x_j} \right),$$

$$\epsilon = \nu \overline{\left(\frac{\partial u_i'}{\partial x_j} \right)^2},$$

where: $e' = \frac{1}{2} u_i'^2$. In this derivation the density and temperature effects have been neglected. At the left-hand side of Eq. (7.8) we have again the change of e along a point that moves at the average speed $\overline{u}_i$. The right-hand side of (7.8) consists of three terms that we discuss separately.

P_k: **production** This is the same term as the *deformation work* in Eq. (7.1) for E. We demonstrated there that the deformation work is, in general, a loss term. Here we find the same term, but with the opposite sign. This implies that P_k in the equation for e is a *production* term, or *source* term. In other words, the average flow supplies energy to the turbulence via deformation work. The mechanisms taking care of the energy transfer are the instability processes discussed in Chap. 3. This term relates the gradient in the average flow (i.e., shear) to the turbulent kinetic energy. Therefore, this term is sometimes called *shear production*.

Let us determine the order of magnitude of the production term. Both the gradient $\partial \overline{u}_i / \partial x_j$ and the Reynolds stress $-\overline{u_i' u_j'}$ scale with characteristic length and velocity of the macrostructure $\mathcal{U}$ and $\mathcal{L}$, so for the production term P_k it follows that:

$$P_k = \mathcal{O} \left(\frac{\mathcal{U}^3}{\mathcal{L}} \right). \tag{7.9}$$

This estimate is somewhat obvious, because we couple the physical mechanism responsible for the production of turbulent kinetic energy to the instability processes, which generated the macrostructure.

$T_k + \Pi_k + D_k$: **transport** Here we again have a term in a divergence form. In the previous section we showed that such a term is responsible for redistribution in space. In this case we find three contributions: (i) transport by velocity fluctuations (T_k), (ii) transport by pressure fluctuations (Π_k), and (iii) transport by viscosity (D_k). A scaling of these three transport terms gives:

$$T_k = \mathcal{O}\left(\frac{U^3}{\mathcal{L}}\right), \quad \Pi_k = \mathcal{O}\left(\frac{U^3}{\mathcal{L}}\right), \quad D_k = \mathcal{O}\left(\frac{U^2}{\mathcal{L}^2}\right).$$

From this it follows that, for large Reynolds numbers, the viscous transport D_k is negligible compared to the other transport terms T_k and Π_k.

Another aspect of the transport term is that we again have a closure problem. The transport term is of a higher order than e, and thus it forms a new unknown variable in the equation for e. On the basis of the same arguments as used in Chap. 5, in relation to (5.26), we can formulate the following closure hypothesis:

$$\overline{u_j'e'} + \frac{1}{\rho_0}\overline{p'u_j'} = -\frac{K}{\sigma_k}\frac{\partial e}{\partial x_j}, \tag{7.10}$$

where σ_k is a constant that is often taken equal to unity. This means that the flux of turbulent kinetic energy is taken proportional to the gradient of the kinetic energy. We can use the K-theory again, with all the drawbacks associated with it, as discussed in Chap. 5. However, in many flows, the transport of e is negligible, so fortunately any problems related to the closure hypothesis in (7.10) play only a minor role.

ϵ: **dissipation** The last term is by definition always negative, and can be considered a loss term, or *sink*, for turbulent kinetic energy. How should we interpret this term? Equation (7.8) is an equilibrium equation for e. Let us limit ourselves to the situation with $De/Dt = 0$, that is, the turbulence is in an *equilibrium*. An equilibrium is only possible when the dissipation is of the same order of magnitude as the production. This holds because the transport term cannot produce or destroy any kinetic energy. Based on estimates of the order of magnitude of the production term P_k, as discussed above, it follows now that:

$$\epsilon = \mathcal{O}\left(\frac{U^3}{\mathcal{L}}\right). \tag{7.11}$$

In other words, we find the result that we postulated as the Kolmogorov relation in Sect. 4.2; here we just demonstrated that this result is based on the balance:

$$\textbf{production} \approx \textbf{dissipation}. \tag{7.12}$$

This result can be considered as one of the cornerstones of turbulence. Of course, Eq. (7.11) does not give the complete picture; from the definition of ϵ in (7.8) it follows that the viscosity ν appears in the dissipation term, while this is absent in (7.11). Therefore, let us scale the dissipation term in (7.8) directly using the viscosity. In this case it follows that:

$$\epsilon = \mathcal{O}\left(\nu \frac{\mathcal{U}^2}{\lambda^2}\right),\tag{7.13}$$

where we used the velocity scale $\mathcal{U}$ as a measure for the velocity fluctuations. For the length scale we introduced a new scale λ. By applying the balance in (7.12) we find that:

$$\frac{\lambda}{\mathcal{L}} \sim Re^{-1/2},\tag{7.14}$$

where the Reynolds number is defined as: $Re = \mathcal{U}\mathcal{L}/\nu$. The length scale λ is thus much smaller than the length scale $\mathcal{L}$ of the macrostructure. The term λ is known as the so-called *Taylor microscale*.

What can be concluded from the relation above? It seems that we need a small length scale to satisfy the balance between production and dissipation in (7.12). Also, we know that the dissipation of turbulent kinetic energy eventually has to take place through viscosity. Together this leads to the conclusion that viscosity can *only* be effective at small scales, because large gradients can be generated at small scales only. When these *large* gradients are multiplied with the *small* viscosity ν (given that we consider a flow at high Reynolds number), this leads to *finite* dissipation rates that can have the same order of magnitude as the production term. In other words, the microstructure is necessary for the dissipation to occur. This is in agreement with the picture we sketched in Chap. 4 in the context of the Burgers equation and the phenomenology of turbulence: viscous dissipation occurs at the microscale.

The Taylor microscale should not be interpreted as a length scale for the smallest eddies in the flow. For this we derived the Kolmogorov length scale η in Chap. 4. The apparent contradiction is found in the fact that it is not allowed to use the velocity scale $\mathcal{U}$ for the microstructure, as we did in (7.13); instead, we should use the Kolmogorov velocity scale υ. Physically speaking the most correct scaling of ϵ is then

$$\epsilon = \nu \frac{\upsilon^2}{\eta^2}.$$

From this, a relation between λ and η can be derived. We return to the Taylor microscale in Chap. 9.

Let us summarize the most important results. Turbulent kinetic energy is produced in the macrostructure as a result of a hydrodynamic instability processes. The macrostructure loses its energy according to (7.11). Via the cascade process this

Fig. 7.2 Terms in the equation for e for a channel flow: $\times$ gradient production P_k; $\square$ dissipation ϵ; $\triangle$ transport by velocity fluctuations T_k; o transport by pressure fluctuations Π_k; $+$ viscous transport D_k. The Reynolds number of the flow is: $Re_* (= Hu_*/\nu) = 180$. From: Mansour et al. (1988)

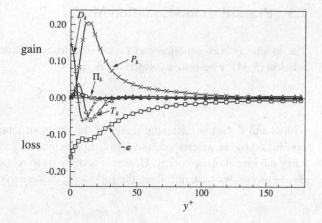

energy eventually ends up at the microstructure, where it is transferred into heat by viscosity.

As an illustration we consider the energy balance for a channel flow. Equation (7.8) can then be simplified to:

$$0 = -\overline{u'v'}\frac{\partial \overline{u}}{\partial y} - \frac{\partial}{\partial y}\left(\overline{v'e'} + \frac{1}{\rho_0}\overline{p'v'} + \nu\frac{\partial e}{\partial y}\right) - \epsilon. \qquad (7.15)$$

The terms in this equation are illustrated in Fig. 7.2 as a function of the dimensionless distance from the wall. These results are based on direct numerical simulations of turbulence in a channel with a relatively low Reynolds number. We find that the largest production occurs in the buffer layer and that the balance between production (P_k) and dissipation (ϵ) is approximately valid over the major part of the channel.

Problems

1. Derive that for the Taylor microscale, defined in (7.13), it holds that: $\lambda/\mathcal{L} = Re^{-1/2}$, where the Reynolds number is defined as: $Re = \mathcal{U}\mathcal{L}/\nu$. Compare this with the ratio $\eta/\mathcal{L}$, where η is the Kolmogorov length scale. Explain this difference by calculating the ratio η/λ.

2. Estimate the characteristic velocity of an eddy with a size λ equal to the Taylor microscale. Show on the basis of this estimate that the energy loss of eddies with a size λ by *direct* viscous dissipation is small.

7.3   Prandtl's One-Equation Model

The kinetic energy equation can be utilized to model turbulent flow. We return to relation (5.34), where we assumed for K that:

$$K \sim \mathcal{U}\mathcal{L},$$

with $\mathcal{U}$ and $\mathcal{L}$ the characteristic scales of the macrostructure. The kinetic energy e is dominated by the macrostructure. This implies that the large eddies are the ones that carry the most kinetic energy. Hence, it is obvious to relate the velocity scale $\mathcal{U}$ to e. Based on this we now introduce the following closure hypothesis:

$$K = c'_\mu \sqrt{e}\mathcal{L}, \tag{7.16}$$

where c'_μ is a constant. The turbulent kinetic energy e is determined by (7.8), which can be written, using the closure hypothesis in (7.10), as:

$$\frac{De}{Dt} = -\overline{u'_i u'_j}\frac{\partial \overline{u}_i}{\partial x_j} + \frac{\partial}{\partial x_j}\left(\frac{K}{\sigma_k}\frac{\partial e}{\partial x_j}\right) - c_D\frac{e^{3/2}}{\mathcal{L}}. \tag{7.17}$$

Here we used the Kolmogorov relation (4.14) with the constant c_D as a closure hypothesis for the dissipation term. For given mixing length $\mathcal{L}$, in combination with (5.20) and (5.26), we have a closed system of equations. This is known as *Prandtl's one-equation model*.

Let us, as a first approximation of Eq. (7.17), neglect the left-hand side and also the transport term at the right-hand side of this equation. Thus, we assume that the turbulence is in an equilibrium state and that turbulent transport is negligible. The equation then reduces to an *exact* balance between production and dissipation. Using (5.26), we find that:

$$K = \sqrt{\frac{c'^3_\mu}{2c_D}\mathcal{L}^2\left|\frac{\partial \overline{u}_i}{\partial x_j} + \frac{\partial \overline{u}_j}{\partial x_i}\right|}. \tag{7.18}$$

This equation is commonly known as the *Smagorinsky model*. It is congruous with the Prandtl mixing length hypothesis, which we discussed in Chap. 5. Hence, Prandtl's one-equation model can be considered as a generalization of the Prandtl mixing length hypothesis.

In (7.16) and (7.17) the constants c'_μ, c_D and σ_k are still unknown, in addition to the length scale $\mathcal{L}$. The constant σ_k is often chosen equal to unity, while the values of the other constants are determined by calibrating the model to a canonical flow, such as the logarithmic layer in a wall-bounded turbulent flow. (Recall that the logarithmic layer is considered to be *universal* to all wall-bounded turbulent flows.) In this logarithmic layer it holds that:

$$u_*^2 \cong -\overline{u'v'} = c_\mu' \sqrt{e} \mathcal{L} \frac{\partial \overline{u}}{\partial y},$$

and the energy equation (7.17) reduces to:

$$u_*^2 \frac{\partial \overline{u}}{\partial y} = c_D \frac{e^{3/2}}{\mathcal{L}},$$

where the transport terms have been neglected on the basis of Fig. 7.2. Given the logarithmic velocity profile (6.15), it follows that:

$$c_\mu' c_D = \frac{u_*^4}{e^2},$$

$$c_\mu' \mathcal{L} = k \frac{u_*}{e^{1/2}} y.$$

On the basis of experimental data and simulation data it follows that: $e/u_*^2 \approx 3.7$ (see Fig. 7.3), and if we further take $\mathcal{L} = ky$, we find that:

$$c_\mu' = 0.52, \quad \text{and:} \quad c_D = 0.14.$$

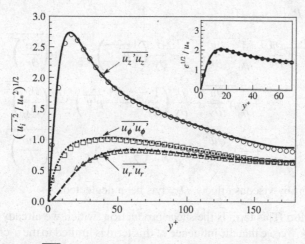

Fig. 7.3 The profiles of $\overline{u_i' u_i'}$ for turbulent pipe flow, where i represents the axial (x), radial (r), and circumferential (ϕ) directions, respectively. The Reynolds number is $Re = 5,300$, based on the bulk velocity and pipe diameter. The *inset* shows the square root $e^{1/2}$ of the kinetic energy relative to u_*. The *lines* represent simulation data from Eggels et al. (1994); the *symbols* represent experimental data from van Doorne and Westerweel (2007)

7.4   Energy Equation per Component

In the previous sections we considered the energy equation summed over the contribution from the three velocity components. In this section we consider the separate contributions from each velocity component. We define:

$$e_\alpha = \frac{1}{2}\overline{u_\alpha'^2},$$

where it is noted that the Greek index letter indicates that no summation is taken over the index. If density and temperature effects are again neglected, we find:

$$\frac{De_\alpha}{Dt} \equiv \frac{\partial e_\alpha}{\partial t} + \overline{u}_j \frac{\partial e_\alpha}{\partial x_j} = -\overline{u_\alpha' u_j'}\frac{\partial \overline{u}_\alpha}{\partial x_j} - \frac{\partial}{\partial x_j}\left(\overline{u_j' e_\alpha} + \frac{1}{\rho_0}\overline{p' u_\alpha'}\delta_{j\alpha}\right) + \frac{1}{\rho_0}\overline{p'\frac{\partial u_\alpha'}{\partial x_\alpha}} - \nu\overline{\left(\frac{\partial u_\alpha'}{\partial x_j}\right)^2}$$

$$(7.19)$$

We can recognize several term from (7.8), except for the fourth term at the right-hand side, which is new. This term is the so-called *pressure–velocity correlation*. For the interpretation of this term we consider the case of a turbulent channel flow in the absence of temperature and density effects, as described in Chap. 6. The equations for the three separate energy components read in this case:

$$
\begin{array}{cccc}
P_{\alpha\alpha} & \Pi_{\alpha\alpha} & T_{\alpha\alpha} & \epsilon_{\alpha\alpha}
\end{array}
$$

$$0 = -\overline{u'v'}\frac{\partial \overline{u}}{\partial y} \quad +\frac{1}{\rho_0}\overline{p'\frac{\partial u'}{\partial x}} \quad -\frac{\partial}{\partial y}\left(\frac{1}{2}\overline{u'^2 v'}\right) \quad -\nu\overline{\left(\frac{\partial u'}{\partial x_j}\right)^2}$$

$$0 = \qquad\qquad +\frac{1}{\rho_0}\overline{p'\frac{\partial v'}{\partial y}} \quad -\frac{\partial}{\partial y}\left(\frac{1}{2}\overline{v'^3} + \frac{1}{\rho_0}\overline{p'v'}\right) \quad -\nu\overline{\left(\frac{\partial v'}{\partial x_j}\right)^2} \qquad (7.20)$$

$$0 = \qquad\qquad +\frac{1}{\rho_0}\overline{p'\frac{\partial w'}{\partial z}} \quad -\frac{\partial}{\partial y}\left(\frac{1}{2}\overline{w'^2 v'}\right) \quad -\nu\overline{\left(\frac{\partial w'}{\partial x_j}\right)^2}$$

where transport by viscous effects, $D_{\alpha\alpha}$ has been neglected.

$P_{\alpha\alpha}$: **production** This term is the shear production, which we already encountered in Eq. (7.8). We see that the influence of this term is limited to the u-component (in the case of plane channel flow). This means that the production of turbulent kinetic energy primarily occurs in a single velocity component. With this, the preferred orientation, or anisotropy, in the turbulence is introduced. This is confirmed by the data presented in Fig. 7.3, from which it follows that fluctuations in the u-component are larger than those in the two other directions. This anisotropy is initially related to the macroscales, because those are directly involved in the processes that lead to the production of turbulent kinetic energy.

$T_{\alpha\alpha}$: **transport** Again, this term represents the transport or redistribution term that we encountered before. This term, which occurs in the three coordinate directions, takes care of a *redistribution* of energy over *space*. We have neglected the viscous contribution to the transport term, using the same reasoning as we used for the transport term T_k in the kinetic energy equation (7.8).

$\epsilon_{\alpha\alpha}$: **dissipation** This is the *dissipation per coordinate-direction*. We saw that the viscous dissipation takes place at the microscales. In a later stage (in particular in Sect. 8.3) we demonstrate that the microstructure is isotropic for large Reynolds numbers. This means that the dissipation per velocity component is equal, and we can therefore say that:

$$\epsilon_{\alpha\alpha} = \frac{1}{3}\epsilon. \tag{7.21}$$

$\Pi_{\alpha\alpha}$: **pressure–velocity correlation** This is the a new term. We saw above that the production of energy at the macroscales is anisotropic, while the dissipation at the microscales is isotropic. During the cascade process the energy thus have to be redistributed over the three coordinate directions. The term $\Pi_{\alpha\alpha}$ is responsible for this, and this is only possible because of the fact that:

$$\sum_{\alpha} \Pi_{\alpha\alpha} = \frac{1}{\rho_0}\overline{p'\left(\frac{\partial u'}{\partial x} + \frac{\partial v'}{\partial y} + \frac{\partial w'}{\partial z}\right)} = 0 \tag{7.22}$$

(for an incompressible flow). Hence, this term, when summed over the three coordinate directions, does not contribute to the total turbulent kinetic energy e. In other words, $\Pi_{\alpha\alpha}$ only achieves a redistribution over the *coordinate directions*. For our example of turbulence in a plane channel flow, where production takes place in the u-direction, we thus expect that: $\Pi_{11} < 0$, with Π_{22} and $\Pi_{33} > 0$. This is confirmed by results from simulations, which that are illustrated in Fig. 7.4

This behavior of $\Pi_{\alpha\alpha}$ suggests a closure hypothesis of the form:

$$\frac{1}{\rho_0}\overline{p'\frac{\partial u'_\alpha}{\partial x_\alpha}} = \frac{\frac{1}{3}e - \frac{1}{2}\overline{u'_\alpha}^2}{\mathcal{T}}, \tag{7.23}$$

which is known as the *Rotta hypothesis*. In (7.23), the term $\mathcal{T}$ is a characteristic time scale that is proportional to $\mathcal{L}/\mathcal{U}$. It can easily be seen that (7.23) satisfies the expression in (7.22).

On the basis of this expression it follows that when an energy component exceeds the isotropic value $\frac{1}{3}e$, the term $\Pi_{\alpha\alpha}$ is a loss term, while in the opposite case $\Pi_{\alpha\alpha}$ is a production term. The turbulence is, so to speak, driven to an isotropic state, and therefore the process represented by these terms is denoted as '*return to isotropy*'.

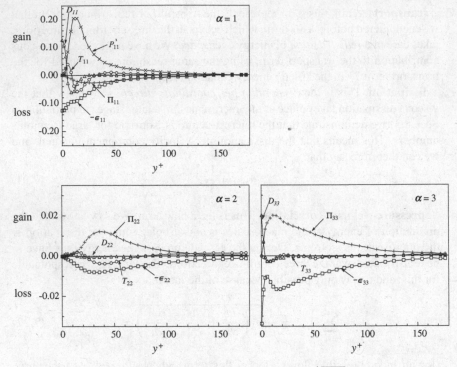

Fig. 7.4 The terms of the energy equation per component $e_\alpha = \frac{1}{2}\overline{u'_\alpha u'_\alpha}$, with $\alpha = 1, 2, 3$, in a channel flow: $\times$ production term $P_{\alpha\alpha}$; $\square$ dissipation $\epsilon_{\alpha\alpha}$; $+$ transport by velocity fluctuations $T_{\alpha\alpha}$; o pressure–velocity correlation $\Pi_{\alpha\alpha}$ (in these graphs $\Pi_{\alpha\alpha}$ is defined as $-\overline{u'_\alpha \partial p'/\partial x_\alpha}$); $\triangle$ transport by viscous diffusion $D_{\alpha\alpha}$. Note that the vertical scale for $\alpha = 1$ is different from those for $\alpha = 2$ and 3. Data from: Mansour et al. (1988)

Problem

1. In a wind tunnel decaying turbulence behind a grid is used for investigating isotropic turbulence. For grid-induced turbulence the velocity fluctuations in the flow direction are larger than in the cross-stream direction: $q_1 > q_2 = q_3$, with: $q_i = \frac{1}{2}\overline{u'_i u'_i}$, where q_i is the kinetic energy (per unit mass) of the i-component of the velocity. The degree of isotropy in the turbulence can be improved by the placement of a contraction directly behind the grid; see Fig. 7.5.

   At the entrance of the contraction the flow has a mean velocity $\overline{u}_1$ along the x_1-axis, and the flow accelerates as it passes through the contraction. The *contraction ratio c* is defined as:

   $$c = \frac{\overline{u}_1(L)}{\overline{u}_1(0)},$$

   where $\overline{u}_1$ is the velocity at the inlet and $\overline{u}_1(L)$ the velocity at the outlet of the contraction.

Fig. 7.5 Flow through a contraction

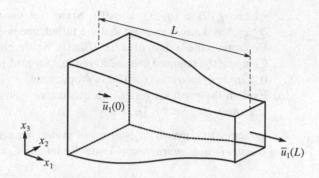

The equation for q_i is:

$$\frac{\partial q_i}{\partial t} + \bar{u}_j \frac{\partial q_i}{\partial x_j} = -\overline{u_i' u_j'} \frac{\partial \bar{u}_i}{\partial x_j} - \frac{\partial}{\partial x_j} \left(\overline{u_j' q_i'} + \frac{1}{\rho} \overline{p' u_i'} \, \delta_{ij} \right) + \frac{1}{\rho} \overline{p' \frac{\partial u_i'}{\partial x_i}} - \nu \overline{\left(\frac{\partial u_i'}{\partial x_j} \right)^2},$$

with: $q_i' = \frac{1}{2} u_i u_i$.

(a) Give the physical meaning of the terms in the equation given above, and identify terms for which a closure relation is required.

(b) Apply this equation to the flow through a contraction; see figure. To satisfy continuity (for an incompressible flow) and for this geometry, the flow near the centerline of the contraction satisfies the following relations:

$$\frac{\partial \bar{u}_1}{\partial x_1} = -2 \frac{\partial \bar{u}_2}{\partial x_2}, \quad \text{and:} \quad \frac{\partial \bar{u}_2}{\partial x_2} = \frac{\partial \bar{u}_3}{\partial x_3},$$

where it is assumed that $\bar{u}_1$ is not a function of x_2 and x_3.
The behavior of the turbulence in the contraction is in close approximation given by the following simplified equations:

$$\bar{u}_1 \frac{\partial q_1}{\partial x_1} = -2q_1 \frac{\partial \bar{u}_1}{\partial x_1}, \quad \bar{u}_1 \frac{\partial q_2}{\partial x_1} = -2q_2 \frac{\partial \bar{u}_2}{\partial x_2}, \quad \bar{u}_1 \frac{\partial q_3}{\partial x_1} = -2q_3 \frac{\partial \bar{u}_3}{\partial x_3},$$

State the arguments for such a simplification that has been applied to the full set of equations for q_i to arrive at the above results for q_1, q_2 and q_3. Under which condition is it acceptable to ignore the dissipation of the turbulent kinetic energy during the passage of the contraction?

(c) Solve the equations for q_1, q_2 and q_3 given above with the boundary conditions at the contraction inlet ($x_1 = 0$):

$$\bar{u}_1 = \bar{u}_1(0), \quad q_1 = q_1(0), \quad q_2 = q_2(0), \quad \text{and:} \quad q_3 = q_3(0),$$

where: $q_1(0) > q_2(0) = q_3(0)$. Show that the total kinetic energy, $e = \sum_i q_i$, has a *minimum*, and that the turbulence is *isotropic* at the minimum. Determine the value of $\bar{u}_1(x_1)/\bar{u}_1(0)$ where this minimum is reached. Compute the required contraction ratio c for grid-generated turbulence with: $0.25q_1 = q_2 = q_3$, to reach an isotropic state.

(d) Explain the previous result in a qualitative manner on the basis of *vortex stretching* (see Sect. 8.1).

A contraction to improve the isotropy of grid-generated turbulence is referred to as a *Comte–Bellot contraction* (Comte–Bellot and Corrsin 1966).

7.5   Convective Turbulence

So far we neglected the influence of variations in density on turbulence. However, in some cases, for example in geophysical flows, density effects play an important role. We can study this using the turbulent kinetic energy equation, where the effect of variations in density and temperature have been taken into account. This means that in the derivation of the kinetic energy equation we have to start with the complete Boussinesq equations given in (5.21) for the velocity fluctuation u_i', including the temperature term. The result then reads:

$$\frac{De}{Dt} = -\overline{u_i'u_j'}\frac{\partial \bar{u}_i}{\partial x_j} + \frac{g}{T_0}\overline{u_j'\theta'}\delta_{j3} - \frac{\partial}{\partial x_j}\left(\overline{u_j'e'} + \frac{1}{\rho_0}\overline{p'u_j'}\right) - \epsilon. \qquad (7.24)$$

In this equation the z-axis is taken vertically, that is, parallel to *gravity*, which is represented by a vector $g_i = (0, 0, -g)$. We see that a new term, $(g/T_0)\overline{w'\theta'}$, appeared. This term is called the *buoyancy production*. The influence of this term is better understood when we couple the temperature flux $\overline{w'\theta'}$ to the temperature gradient using K-theory:

$$-\overline{w'\theta'} = K_H\frac{\partial \bar{\theta}}{\partial z}. \qquad (7.25)$$

We can now distinguish two cases:

$\overline{w'\theta'} > 0$, **or:** $\partial\bar{\theta}/\partial z < 0$

In Chap. 3 we saw that $\partial\bar{\theta}/\partial z < 0$ is an *unstable* situation for small perturbations (that is, a cold (heavy) fluid on top of warm (lighter) fluid). The situation is then characterized by rising ($w' > 0$) and warm ($\theta' > 0$) or descending ($w' < 0$) and cold ($\theta' < 0$) fluid. In both cases we have: $\overline{w'\theta'} > 0$, which is in agreement with (7.25).

On the basis of (7.24) we find that such a temperature flux leads to an *increase* of turbulent kinetic energy. This is consistent with the unstable character of the flow. The velocities caused by the density differences thus directly enhance the turbulent kinetic energy. This is called *buoyant production of turbulence*, as opposed to the production by shear, which we encountered in Sect. 7.2. In a situation where

$(g/T_0)\overline{w'\theta'}$ is the *dominant* source of turbulence, we refer it as *convective turbulence*.

$\overline{w'\theta'} < 0$, **or:** $\partial\overline{\theta}/\partial z > 0$

In Chap. 3 we saw that in this case the flow is *stable* for small perturbations. Indeed, it takes energy to displace fluid elements away from their equilibrium position. In this case the generation of velocity fluctuations requires energy, which is confirmed by (7.24). This is called *buoyant destruction*. In such a situation work has to be done to move the cold ($\theta' < 0$) fluid upwards ($w' > 0$) or the warmer ($\theta' > 0$) fluid downwards ($w' < 0$). This work is done at the expense of the turbulent kinetic energy.

We now have an equation with two production terms: the *shear production* and the *buoyant production*. The ratio between both terms determines the character of the turbulence. This ratio is called the *flux Richardson number*:

$$Ri_f = \frac{\dfrac{g}{T_0}\overline{w'\theta'}}{\overline{u_i'u_j'}\dfrac{\partial\overline{u}_i}{\partial x_j}}. \tag{7.26}$$

We now distinguish three cases:

1. $Ri_f < 0$ In this case the flow is *unstable*, or *convective*. Both the shear and the buoyant production terms are sources of turbulent kinetic energy, and we expect a high turbulence intensity.
2. $Ri_f > 0$ The flow is now called *stable*. In general such a flow has a low turbulence intensity.
3. $Ri_f = 0$ Buoyancy effects are negligible, and only the shear production plays a role. We call this flow neutral.

In the case $Ri_f > 0$ we can expect that, as the stability, or Ri_f, increases, the turbulence eventually *decays*. We can deduce an upper limit for Ri_f on the basis of the following consideration. We neglect the transport term in the energy equation so that:

$$\frac{De}{Dt} = -\overline{u_i'u_j'}\frac{\partial\overline{u}_i}{\partial x_j}\left(1 - Ri_f\right) - \epsilon. \tag{7.27}$$

From this it follows that turbulence always decays (i.e., $De/Dt < 0$) when $Ri_f > 1$. As already mentioned, this is an upper limit, because measurements indicate that turbulence already decays for $Ri_f \sim 0.2$.

In the remainder of the section we focus on turbulence that is dominated by production due to variations in density, which are effectively caused by variations in temperature. The simplest flow geometry in which this kind of turbulence can be studied consists of two flat horizontal plates separated by a distance of $2H$. We already encountered this geometry in Chap. 3 when we discussed the Bénard instability. The lower plate is heated with a constant *temperature flux* $\overline{w'\theta'}_s$. The upper plate is considered to be insulated, so that: $\overline{w'\theta'}(2H) = 0$. Moreover, we assume that the average flow velocity $\overline{u}_i$ equals zero. (With this we eliminate the possibility of turbulence production by shear.)

In such a geometry, we can consider turbulence as *horizontally homogeneous*. On the basis of (5.22) it follows for the equation of the average temperature that

$$\frac{\partial \overline{\theta}}{\partial t} = -\frac{\partial \overline{w'\theta'}}{\partial z}. \tag{7.28}$$

Due to the constant heat input of the lower plate, the average temperature increases continuously as a function of time. However, we can expect that, after damping of possible transient phenomena, the temperature profile attains a time-invariant shape, that is: $\partial \overline{\theta}/\partial z \neq f(t)$. Application of this condition in (7.28) leads to

$$\frac{\partial^2 \overline{w'\theta'}}{\partial z^2} = 0,$$

or equivalently:

$$\overline{w'\theta'} = \overline{w'\theta'}_s \left(1 - \frac{z}{2H}\right). \tag{7.29}$$

The temperature flux profile thus varies linearly with height. This situation, where the average field does depend on time but where the gradients are time-independent, is called: quasi-stationary. This means that the turbulence in particular is stationary, because the production terms of turbulent kinetic energy generally relate to the *gradients* of the average field.

Let us now consider the characteristic scaling of this flow as we applied it for example in Sect. 6.3. It is clear that in the center region, or *core region*, of the channel, H represents the *characteristic length scale*. But, how do we select an appropriate *velocity scale*? For this we use the kinetic energy equation (7.24) for stationary and horizontally homogeneous turbulence. We integrate this energy equation between both plates, where the contribution of the transport terms is zero by definition. It then follows that:

$$0 = \int_0^{2H} \left(\frac{g}{T_0}\overline{w'\theta'} - \epsilon\right) dz.$$

For the term $\overline{w'\theta'}$ we use the profile (7.29), and for ϵ we use the Kolomogorov relation (4.14) with $\mathcal{L} \sim H$. Then it follows for the characteristic velocity scale $\mathcal{U}$ that

$$w_* \equiv \mathcal{U} = \left(\frac{g}{T_0}\overline{w'\theta'}_s H\right)^{1/3}. \tag{7.30}$$

The term w_* is called the *convective velocity scale*. Finally, we can introduce a *convective temperature scale* T_* as

$$T_* = \frac{\overline{w'\theta'}_s}{w_*} \tag{7.31}$$

Following to the same procedure as we used in Sect. 6.3, we can now define scaling laws for the core region, or *outer region*, of our channel; for example,

$$\overline{w'^2} = w_*^2 f_w\left(\frac{z}{H}\right), \quad \overline{u'^2} = w_*^2 f_u\left(\frac{z}{H}\right), \quad \overline{\theta'^2} = T_*^2 f_\theta\left(\frac{z}{H}\right) \tag{7.32}$$

Such scaling is known as *convective scaling*. Let us apply our convective scaling to the average temperature profile. It follows that:

$$\frac{\overline{\theta} - T_{cl}}{T_*} = F(\zeta), \quad \text{with: } \zeta = \frac{z}{H}, \tag{7.33}$$

for the scaling law in the outer region, where T_{cl} is the temperature in the center of the channel, i.e. $z = H$. Close to either of the plates, that is in the *inner region*, we have to consider a different length scale, which describes the characteristics of the plate. This length scale is denoted as z_H. We still expect that the temperature flux $\overline{w'\theta'}_s$ dominates the turbulence. It then follows for the velocity scale in the inner region that:

$$w_f = \left(\frac{g}{T_0}\overline{w'\theta'}_s z_H\right)^{1/3}, \tag{7.34}$$

so that the characteristic scale for the temperature becomes:

$$T_f = \frac{\overline{w'\theta'}_s}{w_f}. \tag{7.35}$$

Now, what would be a suitable choice for z_H? For a smooth plate, z_H should relate to the molecular conduction that takes care of the temperature flux in the proximity of the plate, where the turbulence has vanished (analogous to the viscous sublayer). It follows that:

$$z_H = \mathcal{O}\left(\frac{\alpha}{w_f}\right).$$

With this we find for w_f:

$$w_f = \left(\frac{g}{T_0}\overline{w'\theta'}_s \alpha\right)^{1/4}.$$

Another extreme case is when $z_H = $ constant. This could for example be the asymptotic limit for a very rough surface. However, this situation is not as straightforward as for the roughness length z_0 in turbulent shear flow, which reaches a constant value. The reason that the behavior of z_H is more complicated, is that momentum transport is effected both by molecular transport and by buoyancy. However, near the wall heat can only be transported by molecular effects, so it is not likely that z_H becomes independent of α. Anyway, for the temperature profile in the inner region we can formulate the following scaling law:

$$\frac{T_s - \bar{\theta}}{T_f} = f(z_*) \quad \text{with:} \quad z_* = \frac{z}{z_H}, \tag{7.36}$$

where T_s is the wall temperature. On the basis of symmetry it follows that

$$\Delta T = 2(T_s - T_{cl})$$

where ΔT represents the temperature difference between the lower and the upper plates. By *matching* the profiles in the overlap area between the inner region and the outer region, we can now derive an explicit expression for the temperature profile. Here we use a matching condition that is similar to the one formulated in Sect. 6.3:

$$\lim_{z/H \to 0} \left(\frac{\partial \bar{\theta}}{\partial z}\right)_{\text{outer}} = \lim_{z/z_H \to \infty} \left(\frac{\partial \bar{\theta}}{\partial z}\right)_{\text{inner}}.$$

By substitution of (7.33) and (7.36) in this relationship and after some straightforward manipulation, using the definitions (7.31) and (7.35) for T_* and T_f, respectively, we find:

$$\zeta^{4/3} \frac{\partial F}{\partial \zeta} = -z_*^{4/3} \frac{\partial f}{\partial z_*} = \text{constant},$$

where ζ is defined in (7.33) and z_* in (7.36). This result appears to be independent of the choice for z_H, as discussed above. Integration of the equations above shows that in the overlap area the temperature varies with z as:

$$\bar{\theta} \sim z^{-1/3}.$$

Also, on basis of these matching results we can find an expression for the relation between ΔT and $\overline{w'\theta'}_s$. For this we follow a procedure similar to the one we used in Sect. 6.3 to derive the friction law in (6.30). The found relation is often expressed as a *Nusselt number Nu*, defined as:

$$Nu = \frac{\overline{w'\theta'}_s 2H}{\alpha \Delta T}.$$

We now find for a smooth wall under the condition $z_H \ll H$:

$$Nu = c\,(Ra\,Pr)^{1/3}, \tag{7.37}$$

where the *Rayleigh number Ra* is defined as:

$$Ra = \frac{\frac{g}{T_0}\Delta T (2H)^3}{\nu \alpha}$$

and the *Prandtl number Pr* as:

$$Pr = \frac{\nu}{\alpha}.$$

Especially the dependence of Nu as a function of the Rayleigh number, given by (7.37), has been confirmed by experimental data. An empirical relation based on observations reads:

$$Nu = 0.069\, Ra^{1/3} Pr^{0.074}.$$

Problems

1. Derive the energy equation per component for turbulent flow, including density effects. Show that the buoyancy production only contributes to the vertical component. Interpret the result. (See also Problem 1 of Sect. 7.6.)
2. Define the average potential energy (per unit mass) in the Boussinesq approximation as:

$$P_e = \frac{1}{\rho_0} \int_{z_1}^{z_2} \overline{\rho}\, g\, z\, dz = -\frac{g}{T_0} \int_{z_1}^{z_2} z \overline{\theta}\, dz,$$

where z_1 and z_2 are the boundaries of the turbulent region. Furthermore, assume that the turbulence is horizontally homogeneous ($\partial/\partial x = \partial/\partial y = 0$). Derive on the basis of Eq. (5.22) for $\overline{\theta}$ (neglecting the molecular conduction term) that:

$$\frac{\partial P_e}{\partial t} = -\frac{g}{T_0} \int_{z_1}^{z_2} \overline{w'\theta'}\, dz.$$

Assuming (7.8), it follows for turbulence *with* density effects that:

$$\frac{\partial E_e}{\partial t} = \frac{g}{T_0} \int_{z_1}^{z_2} \overline{w'\theta'}\, dz - \int_{z_1}^{z_2} \epsilon\, dz,$$

with:

$$E_e = \int_{z_1}^{z_2} e\, dz.$$

From this it follows that the term $(g/T_0)\,\overline{w'\theta'}$ can be considered as the rate of conversion for potential energy into kinetic energy.

3. Consider a one-dimensional flow: $\bar{u}_i = (\bar{u}(z), 0, 0)$. On the basis of K-theory, derive a relation between the flux Richardson number Ri_f and the Richardson number Ri, defined as:

$$Ri = \frac{\dfrac{g}{T_0}\dfrac{\partial\bar{\theta}}{\partial z}}{\left(\dfrac{\partial\bar{u}}{\partial z}\right)^2}.$$

With this, estimate a critical value for the flux Richardson number. (In Problem 2 of Sect. 3.3 we found a critical value of $\frac{1}{4}$ for the Richardson number.)

4. Expand Prandtl's one-equation model described in Sect. 7.3 to include density effects. With this, generalize the Smagorinsky model in (7.18) to:

$$K = \sqrt{\frac{c'_\mu{}^3}{2c_D}\mathcal{L}^2}\left|\frac{\partial u_i}{\partial x_j} + \frac{\partial u_j}{\partial x_i}\right|\left(1 - Ri_f\right)^{1/2}.$$

Also, derive that in this case the following holds:

$$e = \frac{\left(1 - Ri_f\right)^{1/2}}{\sqrt{2c'_\mu c_D}}\left(\tau_{ij}^{(t)}\tau_{ij}^{(t)}\right)^{1/2},$$

where: $\tau_{ij}^{(t)} = -\overline{u'_i u'_j} + \frac{1}{3}\overline{u'_k{}^2}\delta_{ij}$ represents the *deviatoric turbulent stress*.

5. Derive the following equation for the temperature fluctuations in a turbulent flow:

$$\frac{D}{Dt}\frac{1}{2}\overline{\theta'^2} \equiv \frac{\partial}{\partial t}\frac{1}{2}\overline{\theta'^2} + \bar{u}_j\frac{\partial}{\partial x_j}\frac{1}{2}\overline{\theta'^2} = -\overline{u'_j\theta'}\frac{\partial\bar{\theta}}{\partial x_j} - \frac{\partial}{\partial x_j}\overline{u'_j\frac{1}{2}\theta'^2} - N,$$

where $N = \alpha\overline{\left(\partial\theta'/\partial x_j\right)^2}$ is called the molecular destruction of temperature fluctuations. The term α is the molecular diffusion coefficient for temperature.

(a) Interpret each of the terms in this equation, and give an estimation of the order of magnitude for each of them.

(b) Introduce the Taylor microscale for temperature fluctuations, and compare this to the Taylor microscale for velocity fluctuations λ.

(c) Derive expressions for the microscales of the temperature fluctuations as a function of the Reynolds number and the Prandtl number. (Doing this, use the Corrsin scales derived in Problem 2 of Sect. 10.2.)

6. Derive relation (7.37) for a smooth wall. Show that for a 'rough' wall, that is $z_H \sim$ constant) this relation changes to:

$$Nu \sim (Ra\,Pr)^{1/2}.$$

7. Consider the convective scaling relations in (7.32).

   (a) Match the expressions for $\overline{w'^2}$ and $\overline{\theta'^2}$ with the relations for these variables as given in Appendix B.1 by application of Monin–Obukhov similarity. With this, derive the expressions found in Problem 3 of Appendix B.1.

   (b) In Appendix B.1 it is stated that $\overline{u'^2}$ does not satisfy the Monin–Obukhov similarity. It appears that this variable meets the convective scaling and that this scaling relation remains valid into the surface layer. With this, derive an expression for $\overline{u'^2}/u_*^2$ in the surface layer.

   (c) Suppose that in the surface layer it holds that:

   $$\frac{\overline{\theta} - T_s}{\theta_*} = f\left(\frac{z}{L}\right),$$

   where θ_* is given by (B.4) and L by (B.3). Match this profile to (7.33), and derive an expression for the Nusselt number Nu.

8. Consider convection between two flat horizontal plates separated by a distance $2H$. The lower plate has a temperature T_s and the upper plate has a temperature T_1, with $\Delta T = T_s - T_1$. After a certain time a stationary situation develops, that is: $\overline{\theta} \neq f(t)$.

   (a) Calculate the temperature flux profile.

   (b) Suppose that in the outer region the characteristic scales are: H and ΔT, and in the inner region: z_H and ΔT. Use matching in order to determine the temperature gradient in the overlap area.

   (c) Suppose that the scaling of turbulent quantities for this flow must yield the same results as the convective scaling discussed previously. What conditions must be satisfied by the Nusselt number Nu?

   (d) It follows from the results for (b) and (c) that, using the scaling as proposed in this problem, results are found that differ from those found using convective scaling. This seems inconsistent. Give a reason why the scaling proposed here is *incorrect* as opposed to convective scaling.

7.6 The Convective Boundary Layer

We now continue our discussion on convective turbulence with a somewhat more complex geometry, that is the *convective atmospheric boundary layer*. The difference with the flow between two flat plates, as discussed in the previous section, is that the upper plate has been removed. In the atmosphere the upper confinement is formed by a *free surface*, which forms the interface between the turbulent boundary layer close to the earth's surface and the free (laminar) atmosphere above this boundary layer.

The average temperature profile in this convective boundary layer is illustrated schematically in Fig. 7.6. Note that in atmospheric flows we often use the potential temperature (see Sect. 2.2), but the equations that describe the flow remain the same. We now indicate the thickness of the boundary layer with z_i (which thus replaces H that we used above). In the largest part of the boundary layer the average temperature θ_{bl} is constant due to the strong mixing that occurs because of the convective turbulence. That is why this layer is referred to as the *atmospheric mixing layer*.

When approaching the surface ($z \to 0$) the temperature in the *surface layer* increases significantly until it reaches the *surface temperature T_s*. The thickness of this surface layer is often small compared to the thickness of the complete boundary layer, and therefore it has been neglected in Fig. 7.6.

At the upper side of the boundary layer at $z = z_i$ we find a temperature jump Δ, which characterizes the free surface as discussed above. In the atmosphere above the boundary layer the temperature generally rises with height. Such an increase is called an *inversion*. We assume that in our case the temperature gradient is constant here, which we indicate by Γ.

Turbulence is produced because the lower wall (i.e., the earth's surface) is heated, in this case by the sun. We use approach of the quasi-stationary turbulence, that is the temperature *gradient* is independent of time, so that the temperature flux is *linear*. At the end of this section we justify this approach.

A difference with convective turbulence between two parallel plates, described in Sect. 7.5, is that in this case the temperature flux (or, actually *heat flux*) at the upper side of the boundary layer ($z=z_i$) does not equal zero. In general we find here a *negative* temperature flux: $\overline{w'\theta'}_i$, as illustrated in Fig. 7.6. We see later that this negative temperature flux relates to the growth of the boundary layer due to *entrainment*, that is air is captured from the free atmosphere into the boundary layer. Because of this entrainment, the thickness of the boundary layer increases as a function of time.

On the basis of the properties discussed above, that is $\theta_{bl} = constant$ and the linearity of $\overline{w'\theta'}$, it follows for the temperature equation that

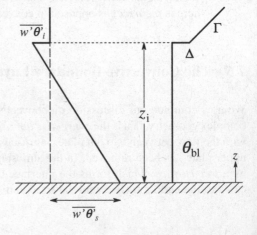

Fig. 7.6 Average temperature and temperature flux as a function of the height in the atmospheric convective boundary layer. The symbols are explained in the text

$$\frac{d\theta_{bl}}{dt} = \frac{\overline{w'\theta'}_s - \overline{w'\theta'}_i}{z_i}. \tag{7.38}$$

Next we integrate the temperature equation (7.28) over the temperature jump at $z = z_i$:

$$\lim_{\delta \to 0} \left(\int_{z_i(t)-\delta}^{z_i(t)+\delta} \frac{\partial \overline{\theta}}{\partial t} \, dz = -\overline{w'\theta'} \Big|_{z_i(t)-\delta}^{z_i(t)+\delta} \right),$$

with:

$$\int_{z_i(t)-\delta}^{z_i(t)+\delta} \frac{\partial \overline{\theta}}{\partial t} = \frac{\partial}{\partial t} \left(\int_{z_i(t)-\delta}^{z_i(t)+\delta} \overline{\theta} \, dz \right) - [\theta(z_i + \delta) - \theta(z_i - \delta)] \frac{dz_i}{dt}.$$

Then, it follows for this limit that:

$$\Delta \frac{dz_i}{dt} = -\overline{w'\theta'}_i, \tag{7.39}$$

which indicates the relation between $\overline{w'\theta'}_i$ and the growth rate of the boundary layer. The last equation is found by differentiating the temperature profile for the free atmosphere:

$$\theta = \theta_{bl} + \Delta + \Gamma \cdot (z - z_i).$$

Because in the free atmosphere we have: $\overline{w'\theta'} = 0$, it follows that: $\partial \overline{\theta}/\partial t = 0$, so that:

$$\frac{d\Delta}{dt} = \Gamma \frac{dz_i}{dt} - \frac{d\theta_{bl}}{dt}. \tag{7.40}$$

We now have a complete set of equations to describe the convective boundary layer. However, we have three equations and four unknown variables. When $\overline{w'\theta'}_s$ and Γ are assumed to be given, the remaining unknown variables are: θ_{bl}, z_i, Δ and $\overline{w'\theta'}_i$. Again, we encounter a closure problem. We can solve this by formulating a closure hypothesis for $\overline{w'\theta'}_i$. For this, we use the equation for turbulent kinetic energy in (7.24). Application of this equation at $z = z_i$ yields:

$$0 = \frac{g}{T_0} \overline{w'\theta'}_i + T_k,$$

where we neglected the dissipation. Given that: $\overline{w'\theta'}_i < 0$, kinetic energy is thus lost. The equation above states that the energy is supplied by the transport term T_k. On the basis of convective scaling we can estimate this transport term as follows:

$$T_k = \mathcal{O} \left(\frac{w_*^3}{z_i} \right),$$

so that, with definition (7.30), it follows that:

$$\overline{w'\theta'}_i = -c\,\overline{w'\theta'}_s,\tag{7.41}$$

where on the basis of experimental data the value for the constant c is found to be 0.1–0.2.

We can now solve the system of equations in (7.38–7.41). For this we first formulate the equation

$$\frac{d\Delta}{dz_i} = \Gamma - \frac{1+c}{c}\frac{\Delta}{z_i}.$$

The solution for this reads:

$$\Delta = \frac{c}{1+2c}\Gamma z_i + A z_i^{-\frac{1+c}{c}},\tag{7.42}$$

where A is an integration constant. The second term at on the right-hand side of (7.42) rapidly approaches zero (for $c \approx 0.1$ for increasing z_i. We therefore can neglect this term, so that:

$$\Delta = \frac{c}{1+2c}\Gamma z_i.$$

After substitution in (7.39) and using the closure hypothesis in (7.41), it now follows for the solution for z_i that:

$$z_i^2(t) = z_i^2(0) + 2(1+2c)\frac{\overline{w'\theta'}_s}{\Gamma}t.\tag{7.43}$$

This solution appears to describe the behavior of the convective boundary layer in the atmosphere quite well. When we consider some typical values: $\overline{w'\theta'}_s = 0.1\,\mathrm{Km/s}$, and: $\Gamma = 5\times10^{-3}\,\mathrm{K/m}$, we find through (7.43) that the boundary layer thickness initially grows at about 400 m per hour.

Finally we have to justify why we can assume quasi-stationary turbulence. An obvious condition is that the time scale relating to the change of the boundary layer thickness is much larger than the time scale of the turbulence. In that case the turbulence experiences a 'constant' (or at least very slowly varying) boundary layer thickness. This condition implies:

$$\frac{1}{z_i}\frac{dz_i}{dt} \ll \frac{w_*}{z_i},$$

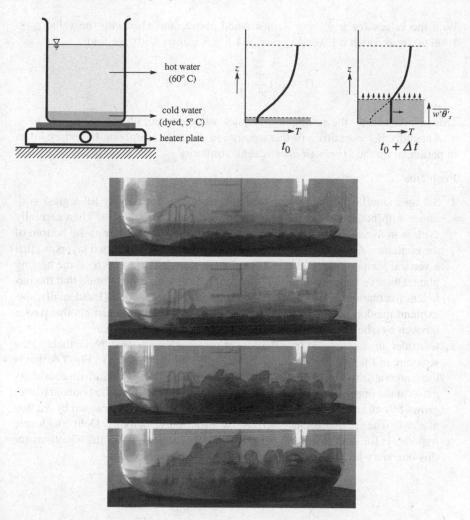

Fig. 7.7 (*Left*) Schematic of a simple demonstration of a convective boundary layer. The preparation of the experiment is described in Problem 1. (*Right*) Initial state and the growth of the convective boundary layer when the heater plate has been switched on. The turbulent mixing in the convective boundary layer results in a uniform temperature. (*Bottom*) Images of the developing turbulent flow region. Turbulent fluctuations are primarily in the vertical direction. The initial growth of the boundary layer thickness z_i follows (7.43)

where we take z_i/w_* for the turbulent time scale. After substitution of (7.43), it follows that:

$$(1 + 2c)\frac{\overline{w'\theta'}_s}{w_* z_i \Gamma} \ll 1.$$

With the values for $\overline{w'\theta'}_s$ and Γ mentioned above, and also with the value $z_i \approx$ 1000 m, from which it follows that $w_* = 1.5$ m/s using (7.30), we find:

$$(1 + 2c)\frac{\overline{w'\theta'}_s}{w_* z_i \Gamma} \approx 0.02 \ll 1.$$

This indeed justifies the assumption of quasi-stationary turbulence.

Appendix B.1 contains a further discussion on the atmospheric boundary layer, in particular on the *Monin-Obukov* scaling similarity.

Problems

1. Set up a small experiment such as depicted in Fig. 7.7. Slowly fill a glass container with hot water (avoid introducing swirl and turbulence). Then carefully syphon in dyed cold water, to form a layer of a few millimeter at the bottom of the container. Allow 1–2 min for the heat transfer between the two layers, so that a vertical temperature profile is formed as depicted. Then switch on the heating plate. Observe the growth of the turbulent region with time. Notice that the turbulent fluctuations are dominated by the vertical component. (Incidentally, this explains the dominant up-and-down motion, or *heaving*, of an airliner that passes through a turbulent atmosphere; see also Problem 1 of Sect. 7.5.)

2. Consider an atmospheric boundary layer for which $\overline{w'\theta'}_s = 0$. Nevertheless, the structure of the boundary layer is comparable to that as shown in Fig. 7.6, that is the temperature in the boundary layer is constant, and a temperature discontinuity exists at the upper side of the boundary layer. Apart from this, the boundary layer grows by entrainment. The turbulence in the boundary layer is caused by *friction* at $z = 0$. The friction is characterized by a friction velocity u_*. Define a closure hypothesis for this situation comparable to (7.41). Next, solve the equations for this boundary layer, and show that the solution is given by:

$$z_i^3(t) = z_i^3(0) + 6c\frac{T_0}{g}\frac{u_*^3}{\Gamma}t.$$

Chapter 8
Vorticity

Vorticity is formally defined as the *curl* of the velocity field:

$$\underline{\omega} = \nabla \times \underline{u}, \quad \text{or:} \quad \omega_i = \epsilon_{ijk}\frac{\partial u_k}{\partial x_j}, \tag{8.1}$$

and can be interpreted as a measure of the rotation of a fluid element. In Chap. 4 we already mentioned the importance of vorticity for turbulence. We even characterized turbulence as 'chaotic vorticity'. In this chapter this is further elaborated.

To get a first notion on the importance of vorticity we rewrite the Reynolds stress as:

$$\frac{\partial \overline{u_i' u_j'}}{\partial x_j} = -\epsilon_{ijk}\,\overline{\omega_k' u_j'} + \frac{\partial}{\partial x_i}\left(\frac{1}{2}\overline{u_j'^2}\right). \tag{8.2}$$

The second term on the right in (8.2) is a gradient of the scalar $\frac{1}{2}\overline{u_j'^2}$, and when we consider the complete equations of motion for $\bar{u}_i$ in (5.20), this term can be added to the pressure gradient. The first term on the right in (8.2) is much more interesting, because it expresses that Reynolds stress implies the existence of *fluctuating* vorticity. We already established the essential role of the Reynolds stress in the preceding chapters. In short, flows void of fluctuating vorticity can not be considered turbulent.

Problem

1. Derive Eq. (8.2). In doing so, make use of

$$\epsilon_{ijk}\,\omega_k' = \frac{\partial u_j'}{\partial x_i} - \frac{\partial u_i'}{\partial x_j}, \quad \text{and:} \quad \frac{\partial u_i' u_j'}{\partial x_i} = u_i'\frac{\partial u_j'}{\partial x_i}$$

(for an *incompressible* flow).

© Springer International Publishing Switzerland 2016
F.T.M. Nieuwstadt et al., *Turbulence*, DOI 10.1007/978-3-319-31599-7_8

8.1   Vorticity Equation

The equation for vorticity can be derived from the Navier–Stokes equations (2.10) by applying the *curl* to both sides of the equation. Clearly, the pressure term vanishes because the curl of a gradient is identical zero. The result reads:

$$\frac{D\omega_i}{Dt} \equiv \frac{\partial \omega_i}{\partial t} + u_j \frac{\partial \omega_i}{\partial x_j} = \omega_j \frac{\partial u_i}{\partial x_j} + \nu \frac{\partial^2 \omega_i}{\partial x_j^2}. \tag{8.3}$$

This equation describes how the vorticity changes as we move along with a fluid element. The second term on the right-hand side of (8.3) is recognized as the *diffusion* of vorticity due to *viscosity*. The first term on the right is new. We rewrite this term as:

$$\omega_j \frac{\partial u_i}{\partial x_j} = \omega_j s_{ij}, \tag{8.4}$$

where

$$s_{ij} = \frac{1}{2} \left(\frac{\partial u_i}{\partial x_j} + \frac{\partial u_j}{\partial x_i} \right)$$

represents the *rate-of-strain tensor* that was introduced in (5.24). The expression in (8.4) thus describes the *interaction* between the fluid *deformation* and the *vorticity*.

The first term on the right-hand side of (8.3), taken together with the complete time derivative on the left-hand side, is identical to the equation for a material *line segment* δX_i in a flow, which reads:

$$\frac{D\,\delta X_i}{Dt} = \delta X_j \, s_{ij}.$$

It is easy to imagine the motion of such a line segment. Under the action of the flow, the length and orientation of the line segment change. This is analogous to the evolution of ω_i; see (8.4). Based on this analogy, we can roughly distinguish two types of interactions, which are coupled to the off-diagonal and diagonal terms respectively of s_{ij}:

$i \neq j$ In this case, s_{ij} describes the angular displacement of a material line segment. With this, (8.4) can be interpreted as a *tilting* of the vorticity vector ω_j from the j-direction to the i-direction. As an illustration we consider a *vortex line* oriented along the x_3-axis, as in Fig. 8.1. In a vortex line all vorticity is concentrated along a line (Kundu and Cohen 2004). Here a vortex line, with components $\omega_j = (0, 0, \omega)$, lies in a flow field with $s_{13} \neq 0$. Because the vortex line (when we ignore viscosity) behaves like a material line segment, it is tilted from the x_3-axis towards the x_1-axis, as depicted in Fig. 8.1. This means that the vorticity vector obtains a non-zero component in the x_1-direction.

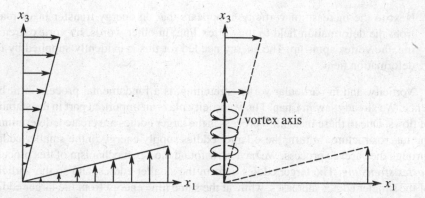

Fig. 8.1 Tilting of a *line vortex* in a deformation field $s_{13} \neq 0$ (The deformation field is indicated in the *left* graph.)

Fig. 8.2 *Vortex line stretching*

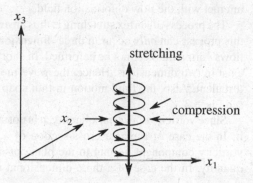

$i = j$ Now, s_{ij} describes the change in *length* of a material line segment, where[1] $s_{\alpha\alpha} > 0$ means that the line segment is stretched, and $s_{\alpha\alpha} < 0$ means that the line segment is compressed (see Problem 1 of Sect. 2.1). Consider for example the situation illustrated in Fig. 8.2, where we show a vortex line $\omega_j = (0, 0, \omega)$ in a velocity field with $s_{33} \neq 0$. When we neglect viscosity, (8.3) reduces to:

$$\frac{D\omega}{Dt} = \omega \, s_{33},$$

with the solution:

$$\omega = \omega(0) \exp(s_{33} \, t).$$

We find that the vorticity increases or decreases depending on the sign of s_{33}. In case $s_{33} > 0$ (i.e., line segments are being stretched) the vorticity increases; we refer to this as *vortex stretching*.

---

[1]Notice that we apply here the convention that we do not sum over the Greek index.

Next to the increase of vorticity, it appears that an energy transfer takes place from the deformation field to the vortex line. In other words, by vortex stretching, the vortex spins up. The energy needed for this is evidently supplied by the deformation field.

Vorticity, and in particular vortex stretching, is a fundamental process in turbulence. We already saw in Chap. 3 how vorticity plays an important part in the stability of flows. Due to these instability processes the larger eddies are created, determining the macrostructure. In turn, these larger eddies supply energy to the smaller eddies through the cascade process. We have now found the key mechanism of this process: *vortex stretching.* The larger eddies deform the smaller eddies, and so the vorticity of the smaller eddies increases, while at the same time energy from the larger eddies is transferred to the smaller eddies. An important consequence of this is that we encounter the largest vorticity magnitude at the microstructure. We come back to this in the sections below; in Appendix B.2 we describe the special case of how vortices interact with the flow deformation field.

The process of vortex stretching is thus an *essential* aspect of turbulence. However, this process can *only* occur in three-dimensional flows, because in two-dimensional flows vortex lines cannot be deformed. In short, turbulence as described here can not exist in two dimensions. Hence, the wave motion at the surface of the ocean is not turbulent. Also, the fluid motion in thin soap films is essentially two-dimensional; see Fig. 8.3.

Since vorticity is a *vector* quantity, it is not straightforward to graphically visualize it. In the case of a planar cross-section of a flow, or a two-dimensional flow, the vorticity component normal to the plane of observation can be plotted as a scalar quantity. In the case of a three-dimensional representation, vortical structures are

Fig. 8.3 The flow behind a grid in a thin soap film. The flow is nearly two-dimensional, and thus there is no energy cascade effective (although some *vortex* pairing may be observed). Compare this to the three-dimensional flow behind a grid in Fig. 4.7. Image courtesy of: M.A. Rutgers (1998)

typically visualized by means of the so-called Q *criterion*. The quantity Q is defined as the *second invariant* of the deformation tensor $\partial u_i / \partial u_j$, defined as (Chong et al. 1990):

$$Q = \frac{1}{2}\left(\mathcal{P}^2 - \frac{\partial u_i}{\partial x_j}\frac{\partial u_j}{\partial x_i}\right), \quad \text{with:} \quad \mathcal{P} = -\frac{\partial u_i}{\partial x_i}. \tag{8.5}$$

Evidently, the first tensor invariant $\mathcal{P} = 0$ for an incompressible flow, so that we can write

$$Q = \frac{1}{2}\left(\frac{1}{2}\omega_i^2 - s_{ij}s_{ji}\right) \tag{8.6}$$

Hence, vortical regions are characterized by large values of Q, so that a region with $Q > Q_E$ identifies vortices or eddies. In practice, an additional criterion for the pressure, i.e. $p < -p_E$, would be required to make a distinction between shear layers and eddies (Hunt et al. 1988).

The *third invariant* $\mathcal{R}$ of the deformation tensor is given by its determinant (Chong et al. 1990)

$$\mathcal{R} = -\det[\partial u_i / \partial x_j], \tag{8.7}$$

which is essentially the product of the tensor eigenvalues. A plot of the Q-$\mathcal{R}$ probability density function shows a characteristic 'teardrop' shape, as shown in Fig. 8.4. The insets in the figure also show the local flow topology identified for the sections of the Q-$\mathcal{R}$ diagram. Note that section I is dominant, which represents an inward spiraling flow in one plane and a strong stretching in a direction away from this plane. This is related to so-called *vortex stretching* discussed at the beginning of the section; this is the dominant small-scale dynamical process in turbulence.

Problems

1. Derive Eq. (8.4).
2. Derive the vorticity equation (8.3). Show that this equation reads in two-dimensional flow:

$$\frac{\partial \omega_i}{\partial t} + u_j \frac{\partial \omega_i}{\partial x_j} = \nu \frac{\partial^2 \omega_i}{\partial x_j^2},$$

with: $\omega_i = (\omega_1, \omega_2)$.

Note that the equation does not include a vortex stretching term. Hence, two-dimensional flows do not have an energy cascade; see Fig. 8.3.

3. Show that the solution of (8.3) when excluding the viscosity term reads:

$$\omega_i(t) = \frac{\delta X_i}{\delta X_j^{(0)}} \omega_j(0),$$

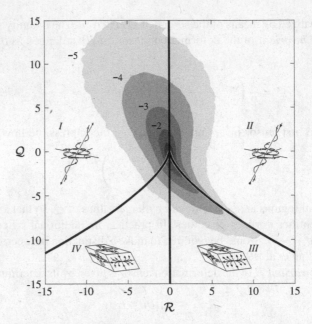

Fig. 8.4 The probability density function (with logarithmic contour intervals) of the second and third invariants, that is $\mathcal{Q}$ and $\mathcal{R}$ respectively, of the deformation tensor. The values of $\mathcal{Q}$ and $\mathcal{R}$ are normalized with respect to the mean of the second invariant of the rotation tensor. The *solid curve* represents a zero value for the discriminant of the deformation tensor. *Insets* represent the local flow topology: I: stable focus/stretching; II: unstable focus/compressing; III: unstable node/saddle/saddle; IV: stable node/saddle/saddle. After: Ooi et al. (1999)

where δX_i represents a material line segment parallel to the vorticity vector $\omega_i(0)$ with a value $\delta X_j^{(0)}$ at time $t = 0$. This solution is known as *Cauchy's equation*. Provide an interpretation for this relationship.

4. Derive Eq. (8.6).

8.2 Coherent Structures

We can use the process of vortex stretching to get a first impression of how turbulent eddies develop. Let us focus on turbulent flow near a wall. We consider a flow with $\overline{u}_i = (\overline{u}(y), 0, 0)$. The deformation coupled to this velocity field consists of a stretching along a line that makes a 45° angle with the wall.

Velocity measurements close to the wall show regions where the velocity is small compared to the direct surroundings. These are called '*low-speed streaks*,' and these are characteristic for all near-wall turbulent flows; see Fig. 8.5. This figure also shows that the low-speed streaks are most visible close to the wall and slowly blur as the distance from the wall increases.

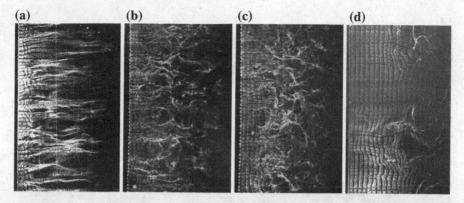

Fig. 8.5 Visualization of the flow in a turbulent boundary layer. The flow is visualized using hydrogen bubbles that are generated by electrolysis from a thin *wire parallel* to the wall, at different distances y (in dimensionless wall units) from the wall. Note the low-speed streaks very close to the wall. Images from: Kline et al. (1967). **a** $y^+ = 2.7$. **b** $y^+ = 38$. **c** $y^+ = 101$. **d** $y^+ = 407$

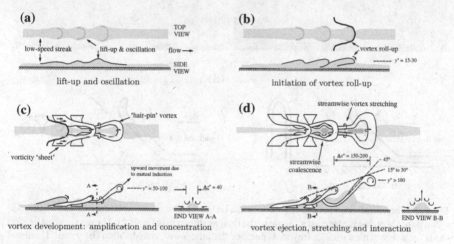

Fig. 8.6 The '*bursting process*' according to Smith (1984). The evolution of *line vortices* in a turbulent flow near a wall: **a** the starting point is a '*low-speed streak*' that becomes perturbed; **b** these perturbations grow because of a Kelvin–Helmholtz instability; **c** the vortex sheet around the low-speed streak rolls up into a *hairpin vortex*; **d** the hairpin vortices are stretched by the mean deformation field until the vortex structure becomes fully unstable and disintegrates in the form of a '*burst*'

The development of a low-speed streak is illustrated schematically in Fig. 8.6. Because the velocity inside a streak is smaller than its surroundings, we can interpret the boundary of the streak as *vortex sheets*.

This flow geometry is comparable to that of the Kelvin–Helmholtz instability, so that any perturbation of these vortex sheets will grow. As a result of the instability the vortex sheet rolls up. The final result is a vorticity distribution concentrated in several so-called *hairpin vortices*. This is illustrated in Fig. 8.6. The tips of these hairpin

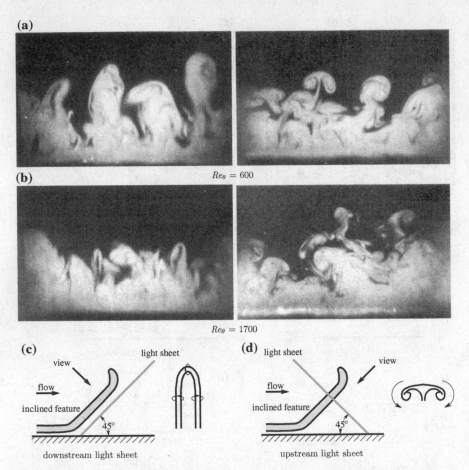

Fig. 8.7 The visualization of hairpin vortices in a turbulent flow near a wall; the flow is visualized in a thin light sheet that is placed at an angle to the flow as shown in (**c**) and (**d**). The *upper* visualization (**a**) is for a flow with a smaller Reynolds number than the *lower* visualization (**b**). From: Head and Bandyopadhyay (1981)

vortices move away from the wall under the influence of the velocity field induced by these vortices. This process is called *self-induction*. This makes the vortex move even further away from the wall into a region with a higher flow speed. The result is that the vortex becomes stretched along the direction of maximal strain, which is along a 45°-line with the wall. Hence, this leads to a physical picture where the boundary layer consists of elongated vortex structures. This is illustrated In Fig. 8.7.

During the stretching process the vorticity grows at the expense of the average flow, until at a certain moment the strength of the hairpin vortex has increased so much that it becomes unstable. The vortex structure then disintegrate into smaller structures. This event is called a turbulent '*burst*.' From measurements it follows that most of the turbulent kinetic energy and turbulent shear stress is produced during burst events.

The hairpin vortex 'pumps' low-speed fluid that is close to the wall into a region away from the wall with a higher mean velocity. This motion is induced by the two counter-rotation vortices that form the 'legs' of the hairpin vortex. This process is referred to as '*ejection.*' It is obvious that this ejection contributes to the Reynolds stress $-\overline{u'v'}$. After the burst, the fluid close to the wall is replaced by high-speed fluid moving towards the wall. This process is called a '*sweep.*' Both ejections and sweeps can be clearly seen in Figs. 5.1 and 5.2, where fluid that moves slower than the (local) mean velocity ($u' < 0$) moves away from the wall ($v' > 0$), and faster fluid ($u' > 0$) moves toward the wall ($v' < 0$).

The complete process, as sketched here, has a cyclic character with a characteristic time scale of $T_B \approx 5\delta/U_0$, where δ is the characteristic scale of the outer region (for example the boundary layer thickness). The term U_0 is the average free-stream velocity. We notice here that this process occurs in a region close to the wall (the complete bursting process takes place in the buffer layer), but it apparently scales with the parameters of the *outer region*. This seems to contradict the scaling principles described in Sect. 6.3. This point has not yet been clarified.

We also saw that vortex stretching plays an important part in the dynamics of these coherent structures. In particular, a preferred direction exists, which is related to the direction of maximum strain, at a 45° downstream angle with the wall. This explains why the large vortices in a turbulent flow are generally anisotropic.

We thus find both structures and processes that can be recognized as such and that are recurrent in a turbulent flow near a wall. These are: the burst, the low-speed streak, the sweep and the ejection. It appears that all these phenomena can be explained in terms of the dynamics of a special vortex structure: the *hairpin vortex* (Adrian 2007). The structures mentioned above, where the hairpin vortex plays a central role, are called *coherent structures*. Coherent structures are also encountered in other types of turbulent flows; for example, the mixing layer, which we discussed in Sect. 6.6 and shown in Figs. 4.6 and 6.16. These coherent structures are considered as an elementary building block for turbulent flow, and for this reason a significant amount of research is dedicated to these structures. An example of an experimental observation is shown in Fig. 8.8. This is an observation by means of PIV of the instantaneous flow field normal to the wall. The structures near the wall can be considered as signatures of the hairpin vortices that occur in the bursting process depicted in Fig. 8.6. These structures have also been identified in state-of-the-art numerical simulations of turbulent flows near a wall (Wu and Moin 2009); see for example Fig. 8.9.

The premise is that when we understand these structures and their dynamics, better closure models for turbulence can be formulated. Perhaps it is even possible to manipulate these structures directly in order to influence turbulent transport processes related to mixing and drag. Examples of the latter one is to achieve a drag reduction by adding polymers to a turbulent flow or by applying small grooves, or *riblets*, to the wall; see also Sect. 6.4.

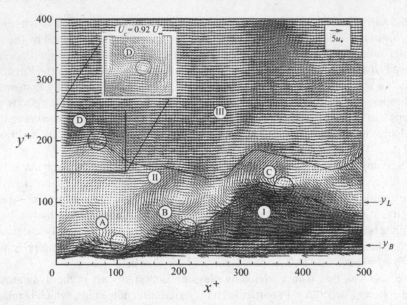

Fig. 8.8 Signatures of hairpin vortices in a PIV measurement in a streamwise-wall-normal plane of a turbulent boundary layer at $Re_\theta = 930$. Labels A–D indicate the heads of the hairpin vortices. Note the correspondence with the flow structure sketched in Fig. 8.6d. The *arrow* plot is generated for a reference frame that is moving at 80 % of the free-stream velocity. From: Adrian et al. (2000)

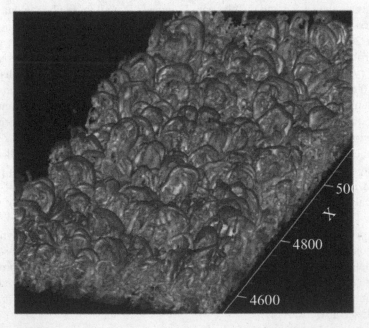

Fig. 8.9 Vortex structures in a downstream section of a zero-pressure-gradient turbulent boundary layer, computed by means of a DNS. The structures are visualized as iso-surfaces of the second invariant $\mathcal{Q}$ of the deformation tensor, where the color indicates the local streamwise velocity. From: Wu and Moin (2009)

Problem

1. Show for a mean flow with a velocity profile $\bar{u}_i = (\bar{u}(y), 0, 0)$ that the direction of maximum strain is at a 45° angle with the flow direction (i.e., x-axis). Hint: find the eigenvalues and eigenvectors of the rate-of-strain tensor s_{ij} near a point $\underline{x}$.

8.3 Enstrophy

In Chap. 7 we studied the equation for the turbulent kinetic energy. Because vorticity plays such an important part in turbulence, it is obvious to consider also the equation for the vorticity fluctuations. We split the instantaneous vorticity ω_i in mean and fluctuating parts, i.e. $\omega_i = \bar{\omega}_i + \omega'_i$. We then define $\frac{1}{2}\overline{\omega'^2_i}$ as the *enstrophy*. How to interpret this new quantity? It can be proven that for homogeneous and isotropic turbulence

$$\epsilon = \nu \overline{\omega'^2_i}. \tag{8.8}$$

This expression is known to be a good approximation for other types of turbulent flows at large Reynolds numbers, for which the turbulence can be considered to be locally isotropic. This relationship means that the enstrophy is dominated by the *microstructure*, where the viscous dissipation occurs. This can also be understood from the fact that ω'_i is directly coupled to $\partial u'_i/\partial x_j$, and we already saw that the largest contribution to the gradient of the velocity fluctuations occurs at the microscale. Hence, we have the turbulent kinetic energy e that is representative of the macrostructure, and the enstrophy $\frac{1}{2}\overline{\omega'^2_i}$ that represents the microstructure.

An equation for the enstrophy can be derived in a similar way as for the turbulent kinetic energy e. In Eq. (8.3) for the vorticity we substitute: $\omega_i = \bar{\omega}_i + \omega'_i$. After averaging this expression we obtain an equation for $\bar{\omega}_i$. This expression is then we subtracted from (8.3), so we arrive at an equation for ω'_i. Finally we multiply this last equation with ω'_i. After averaging, the result reads:

$$\frac{D}{Dt} \frac{1}{2}\overline{\omega'^2_i} \equiv \frac{\partial}{\partial t} \frac{1}{2}\overline{\omega'^2_i} + \bar{u}_j \frac{\partial}{\partial x_j} \frac{1}{2}\overline{\omega'^2_i} = P_\omega + T_\omega + D_\omega + S_\omega - \epsilon_\omega, \tag{8.9}$$

with:

$$P_\omega = -\overline{u'_j \omega'_i} \frac{\partial \bar{\omega}_i}{\partial x_j},$$

$$T_\omega + D_\omega = \frac{\partial}{\partial x_j} \left\{ -\overline{u'_j \frac{1}{2}\omega'^2_i} + \nu \frac{\partial}{\partial x_j} \left(\frac{1}{2}\overline{\omega'^2_i} \right) \right\}$$

$$S_\omega = \overline{\omega'_i \omega'_j s_{ij}} + \overline{\omega'_i \omega'_j} \bar{s}_{ij} + \bar{\omega}_i \overline{\omega'_j s'_{ij}} \tag{8.10}$$

$$\epsilon_\omega = \nu \overline{\left(\frac{\partial \omega'_i}{\partial x_j} \right)^2}.$$

The left-hand side of (8.9) describes how the enstrophy changes for a point moving with the mean flow. The four combined terms on the right-hand side describe the processes that result in a change of enstrophy. These are:

P_ω: **gradient production** We can consider this first term as the gradient production of enstrophy, and as such it is comparable with the gradient production term in the equation for the turbulent kinetic energy. Based on this, it follows that via this term enstrophy is exchanged between the average and the fluctuating vorticity field.

$T_\omega + D_{\omega}$: **enstrophy transport** Due to this term, enstrophy is redistributed in space. We find two contributions: transport by velocity fluctuations (T_ω) and transport by viscosity (D_ω).

S_ω: **stretching** This is the combination of terms relating directly to the process of vortex stretching. We see here both the contribution from the average deformation field $\bar{s}_{ij}$ and from the fluctuating deformation field s'_{ij}.

ϵ_ω: **molecular destruction** This term is always negative, and is thus a loss term. It represents the destruction of enstrophy by molecular diffusion.

We can get a first impression of the various terms in (8.9) on the basis of a direct numerical simulation of a channel flow, of which the results are shown in Fig. 8.10.

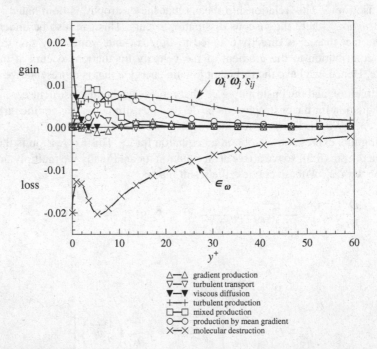

Fig. 8.10 Terms in the enstrophy Eq. (8.9) for a channel flow: $\triangle$ gradient production P_ω; $\triangledown$ transport T_ω and $\blacktriangledown$ viscous diffusion D_ω; stretching terms S_ω, with: $+$ turbulent production $\overline{\omega'_i \omega'_j s'_{ij}}$, $\square$ mixed production $\overline{\bar{\omega}_i \omega'_j s'_{ij}}$, and o production by the mean gradient $\overline{\omega'_i \omega'_j \bar{s}_{ij}}$; and, $\times$ molecular destruction ϵ_ω. After: Mansour et al. (1988) (Note: our notation is somewhat different from the labeling of the original data.)

We see that in the middle of the channel, (8.9) is virtually reduced to an equilibrium between two terms, namely:

$$\overline{\omega_i' \omega_j' s_{ij}'} \approx \epsilon_\omega.$$

This simplification of the enstrophy equation can be justified through a scaling analysis, where the orders of magnitude of the various terms in (8.9) are estimated. We define the following scaling for the average and the fluctuating parts of the vorticity field and the deformation field:

$$\overline{\omega}_i = \mathcal{O}\left(\frac{\mathcal{U}}{\mathcal{L}}\right), \quad \omega_i' = \mathcal{O}\left(\frac{\mathcal{U}}{\lambda}\right), \quad \overline{s}_{ij} = \mathcal{O}\left(\frac{\mathcal{U}}{\mathcal{L}}\right), \quad s_{ij}' = \mathcal{O}\left(\frac{\mathcal{U}}{\lambda}\right), \quad (8.11)$$

where λ is the *Taylor microscale* that we introduced in (7.13). $\mathcal{U}$ and $\mathcal{L}$ are the macroscales, which we already know. Based on this we can scale the terms in (8.9) as follows:

$$
\begin{aligned}
\overline{\omega_i' \omega_j' s_{ij}'} &= \mathcal{O}\left(\frac{\mathcal{U}^2}{\lambda^2}\frac{\mathcal{U}}{\lambda}\right) = \mathcal{O}\left(\frac{\mathcal{U}^3}{\lambda^3}1\right), \\
\frac{\partial}{\partial t}\frac{1}{2}\overline{\omega_i'^2} &= \mathcal{O}\left(\frac{\mathcal{U}}{\mathcal{L}}\frac{\mathcal{U}^2}{\lambda^2}\right) = \mathcal{O}\left(\frac{\mathcal{U}^3}{\lambda^3}\frac{\lambda}{\mathcal{L}}\right), \\
\overline{u}_j\frac{\partial}{\partial x_j}\frac{1}{2}\overline{\omega_i'^2} &= \mathcal{O}\left(\frac{\mathcal{U}}{\mathcal{L}}\frac{\mathcal{U}^2}{\lambda^2}\right) = \mathcal{O}\left(\frac{\mathcal{U}^3}{\lambda^3}\frac{\lambda}{\mathcal{L}}\right), \\
\frac{\partial}{\partial x_j}\overline{u_j'\frac{1}{2}\omega_i'^2} &= \mathcal{O}\left(\frac{\mathcal{U}}{\mathcal{L}}\frac{\mathcal{U}^2}{\lambda^2}\right) = \mathcal{O}\left(\frac{\mathcal{U}^3}{\lambda^3}\frac{\lambda}{\mathcal{L}}\right), \\
\overline{\omega_i' \omega_j'}\,\overline{s}_{ij} &= \mathcal{O}\left(\frac{\mathcal{U}^2}{\lambda^2}\frac{\lambda}{\mathcal{L}}\frac{\mathcal{U}}{\mathcal{L}}\right) = \mathcal{O}\left(\frac{\mathcal{U}^3}{\lambda^3}\frac{\lambda^2}{\mathcal{L}^2}\right), \\
-\overline{u_j'\omega_i'}\frac{\partial\overline{\omega}_i}{\partial x_j} &= \mathcal{O}\left(\frac{\mathcal{U}^2}{\mathcal{L}}\frac{\mathcal{U}}{\mathcal{L}^2}\right) = \mathcal{O}\left(\frac{\mathcal{U}^3}{\lambda^3}\frac{\lambda^3}{\mathcal{L}^3}\right), \\
\overline{\omega}_i\,\overline{\omega_j' s_{ij}'} &= \mathcal{O}\left(\frac{\mathcal{U}}{\mathcal{L}}\frac{\mathcal{U}^2}{\mathcal{L}^2}\right) = \mathcal{O}\left(\frac{\mathcal{U}^3}{\lambda^3}\frac{\lambda^3}{\mathcal{L}^3}\right), \\
\nu\frac{\partial^2}{\partial x_j^2}\left(\frac{1}{2}\overline{\omega_i'^2}\right) &= \mathcal{O}\left(\frac{\nu}{\mathcal{L}^2}\frac{\mathcal{U}^2}{\lambda^2}\right) = \mathcal{O}\left(\frac{\mathcal{U}^3}{\lambda^3}\frac{\lambda^3}{\mathcal{L}^3}\right).
\end{aligned}
\qquad (8.12)
$$

For details of this analysis we refer to the book of Tennekes and Lumley (1972) and to one of the problems at the end of this section.

We have put all terms in increasing order of $\lambda/\mathcal{L}$. We know on the basis of (7.14) that for large Reynolds numbers the ratio $\lambda/\mathcal{L} \to 0$. It then follows that the lowest-order terms in $\lambda/\mathcal{L}$ form a first-order approximation of the enstrophy equation. For the destruction term ϵ_ω we did not make an estimation, but we expect that this term is the dominant loss term, i.e. of the lowest-order in $\lambda/\mathcal{L}$. On this basis of this argument it follows that:

$$\overline{\omega_i' \omega_j' s_{ij}'} = \nu \overline{\left(\frac{\partial \omega_i'}{\partial x_j}\right)^2} \tag{8.13}$$

holds as a first-order approximation of the enstrophy equation in (8.9).

We can interpret this expression as follows. Production of enstrophy by vortex stretching is in equilibrium with molecular destruction. The essence of the equation is, however, that the production and the destruction *both* take place at the microscales, and in (8.13) no terms occur that relate directly to the macrostructure. In other words, the microstructure is *dynamically independent* of the macrostructure; that is, the microstructure and the macrostructure are *decoupled* and only connected indirectly through the cascade process.

The result (8.13) is known as the *local equilibrium of the microstructure*. It is one of the cornerstones of current turbulence theory. In particular, it forms the basis of the Kolmogorov scaling of the microstructure, which we discussed in Sect. 4.2. Also, the absence of a direct influence of the macrostructure in (8.13) means that the microstructure has no preferred direction. In other words, the microstructure is *isotropic*. This confirms our postulate in Sect. 4.2, based on flow visualizations such as in Figs. 4.6, 4.7, and 6.23.

Problems

1. Prove that the relation $\epsilon = \nu \overline{\omega_i'^2}$ is exact for homogeneous and isotropic turbulence at large Reynolds numbers.
2. Show by scaling of the first-order approximation (8.13) that the Kolmogorov length scale is the characteristic length scale for the gradient of vorticity fluctuations.
3. Experimental and numerical data suggest that the dissipation ϵ is not distributed uniformly over the complete turbulent flow; see Fig. 8.11. It is said that the dissipation is *intermittent*, i.e. the value of the dissipation is large in a small volume, while in most of the domain dissipation is negligible.

   Construct a model that assumes that dissipation takes place in thin vortex tubes with a diameter η (i.e., the Kolmogorov length scale) and in which the characteristic velocity is $\mathcal{U} = \sqrt{\frac{2}{3}e}$. What fraction of the volume is occupied by these vortex tubes? Does the model satisfy the simplified enstrophy budget in (8.13)? (Note: one is free to make a choice for the scaling of s_{ij}'.)
4. Derive the equation for $\overline{\omega}_i$. Show that P_ω is the exchange term between the expressions for $\overline{\omega}_i$ and $\frac{1}{2}\overline{\omega_i'^2}$.
5. Show that the enstrophy equation for a fluctuating two-dimensional flow reads

$$\frac{D}{Dt}\frac{1}{2}\overline{\omega_i'^2} = -\overline{u_j'\omega_i'}\frac{\partial \overline{\omega}_i}{\partial x_j} - \frac{\partial}{\partial x_j}\left\{\overline{u_j'\frac{1}{2}\omega_i'^2} - \nu\frac{\partial}{\partial x_j}\left(\frac{1}{2}\overline{\omega_i'^2}\right)\right\} - \nu\overline{\left(\frac{\partial \omega_i'}{\partial x_j}\right)^2}.$$

Compare this expression with (7.8) for e. Based on this equation for e we have given arguments for the cascade process. Argue on the basis of the equation

Fig. 8.11 The three-dimensional distribution of the vorticity magnitude (above a certain threshold) in a simulation of homogeneous and isotropic turbulence. The vorticity magnitude, which is proportional to the enstrophy, is spatially localized in thin worm-like structures. From: Vincent and Meneguzzi (1991)

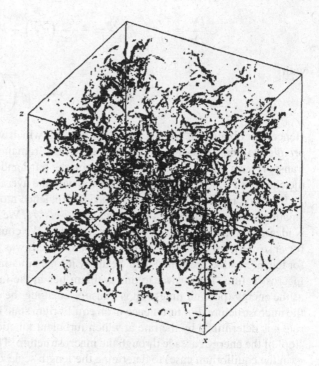

above that for two-dimensional flows an *enstrophy cascade* occurs with ϵ_ω as the destruction of enstrophy (where the term ϵ_ω can be compared with ϵ in the equation for e). What are the microscales in this flow?

6. Show that the term $\overline{\omega_i' \omega_j' s_{ij}'}$ in the enstrophy equation always leads to an *increase* of enstrophy. (Hint: consider this term in a so-called principal coordinate system in which s_{ij}' reduces to a diagonal form $s_{ij}' = s_{ik}' \delta_{jk}$ with $s_{11} + s_{22} + s_{33} = 0$.)

7. Derive the scaling relations in (8.12) using (8.11). When scaling $\overline{\omega_i' \omega_j'}$, we use the fact that the microstructure is isotropic. From this it follows that the correlation between ω_i' and ω_j' is small, that is $\mathcal{O}(\lambda/\mathcal{L})$. For the scaling of the term $\overline{\omega_i' s_{ij}'}$ we use Eq. (8.2). The scaling of $\overline{\omega_i' s_{ij}'}$ follows from Eq. (8.4) and from the fact that the correlation between u_j' and ω_i' is also small, that is $\mathcal{O}(\lambda/\mathcal{L})$. The reason for this is that u_j' is determined by the macrostructure and ω_i' by the microstructure.

8.4 The k-ϵ (e-ϵ) Model

We now consider a second-order approach of the enstrophy equation according to the scaling given in (8.12), i.e. we include all terms down to order $\mathcal{O}(\lambda/\mathcal{L})$. At the same time we substitute: $\epsilon = \nu \overline{\omega_i'^2}$, and: $\epsilon' = \nu \omega_i'^2$, with the result:

$$\frac{D\epsilon}{Dt} \equiv \frac{\partial \epsilon}{\partial t} + \overline{u}_j \frac{\partial \epsilon}{\partial x_j} = -\frac{\partial}{\partial x_j}\left(\overline{u'_j \epsilon'}\right) + P_\epsilon - D_\epsilon \qquad (8.14)$$

with

$$P_\epsilon = 2\nu\,\overline{\omega'_i \omega'_j s'_{ij}}, \quad \text{and:} \quad D_\epsilon = 2\nu^2 \overline{\left(\frac{\partial \omega'_i}{\partial x_j}\right)^2}. \qquad (8.15)$$

Hence, we obtained a balance equation for ϵ in which we can interpret P_ϵ as the *production* and D_ϵ as the *destruction* of ϵ. The remaining terms in (8.14) are a transport term on the right-hand side and on the left-hand side a term that describes the change of ϵ along a point which moves with the average flow speed.

Based on Eq. (8.14) we can formulate a turbulence model. Indeed, using the dissipation ϵ we can introduce a length scale: $\mathcal{L} \sim \mathcal{U}^3/\epsilon$. We notice that this expression is identical to the Kolmogorov relation in (4.14). We could object that the equation for enstrophy, on which the equation for ϵ is based, was introduced as an equation for the microstructure, while here we apply it for estimating the length scale $\mathcal{L}$ of the macrostructure. The answer to this problem lies in the fact that we interpret ϵ here as the energy (per unit time) that is transported along the cascade to be dissipated at the microstructure. For turbulence in an equilibrium state it holds that the dissipation rate ϵ is determined by the rate at which turbulent kinetic energy is supplied at the 'top' of the energy cascade through the macrostructure. This implies that we can use ϵ (in the equilibrium case) to determine the length scale $\mathcal{L}$ of the macrostructure.

For the velocity scale $\mathcal{U}$ we use again the kinetic energy e. This means that we can define both a characteristic velocity scale and a characteristic length scale, so given that $K \sim \mathcal{U}\mathcal{L}$ we can write:

$$K = c_\mu \frac{e^2}{\epsilon}, \qquad (8.16)$$

where c_μ is a constant. Equation (8.16) is known as the k-ϵ *model*, although in view of the notation used in this book it would be more appropriate to call it the e-ϵ model.

In order to apply the k-ϵ model (8.16) we need to specify another equation for e and for ϵ. We have already discussed the equation for e in Sect. 7.2. The equation for ϵ is given above in (8.14), but cannot be used in that particular form. This is because some unknown terms occur in (8.14), for which we need to specify yet another closure hypothesis.

In the previous section we found that P_ϵ and D_ϵ form a first-order equilibrium. The difference between those two terms is of secondary order, and we assume here that this difference is proportional to the production and dissipation terms in the expression for the turbulent kinetic energy, i.e.

$$P_\epsilon - D_\epsilon = \frac{\epsilon}{e}(c_{1\epsilon}P - c_{2\epsilon}\epsilon). \qquad (8.17)$$

Here

$$P = -\overline{u_i' u_j'} \frac{\partial \overline{u}_i}{\partial x_j} + \frac{g}{T_0} \overline{u_j' \theta'} \delta_{j3}$$

is the production of kinetic energy. The terms $c_{1\epsilon}$ and $c_{2\epsilon}$ in (8.17) are unknown constants, and $\mathcal{T} = e/\epsilon$ can be considered as a characteristic time scale. For the transport term we choose the usual gradient hypothesis (or, K-theory):

$$-\overline{u_j' \epsilon'} = \frac{K}{\sigma_\epsilon} \frac{\partial \epsilon}{\partial x_j}, \tag{8.18}$$

where σ_ϵ is a constant. A somewhat more sophisticated closure for the transport term, which takes into account the anisotropy of this term, is given by:

$$-\overline{u_j' \epsilon'} = c_\epsilon \overline{u_k' u_j'} \frac{e}{\epsilon} \frac{\partial \epsilon}{\partial x_k},$$

where c_ϵ is a constant for which a value of 0.15 has been proposed.

With this our k-ϵ model is closed, and for completeness we here summarize the complete set of equations:

$$K = c_\mu \frac{e^2}{\epsilon},$$

$$\frac{D\epsilon}{Dt} = \frac{\partial}{\partial x_j} \left(\frac{K}{\sigma_\epsilon} \frac{\partial \epsilon}{\partial x_j} \right) + \frac{\epsilon}{e}(c_{1\epsilon}P - c_{2\epsilon}\epsilon), \tag{8.19}$$

$$\frac{De}{Dt} = P + \frac{\partial}{\partial x_j} \left(\frac{K}{\sigma_k} \frac{\partial e}{\partial x_j} \right) - \epsilon.$$

In combination with Eqs. (5.20) and (5.26), this forms a closed system of equations. The constants c_μ, $c_{1\epsilon}$, $c_{2\epsilon}$, σ_k, and σ_ϵ in (8.19) are determined on the basis that the k-ϵ model should satisfy some canonical turbulent flows.

We first consider *decaying homogeneous turbulence*. In practice, this type of turbulence is produced by placing a grid in a wind tunnel. In the wake of the grid, turbulence emerges that slowly decays while being advected downstream by the main flow. A visualization of such a flow is shown in Fig. 4.7. In the ideal case of homogeneous and isotropic turbulence *all* spatial gradients of the average flow properties are equal to zero. With this, Eq. (8.19) reduces to:

$$\frac{\partial \epsilon}{\partial t} = -c_{2\epsilon} \frac{\epsilon^2}{e},$$

$$\frac{\partial e}{\partial t} = -\epsilon. \tag{8.20}$$

First, we consider an expression for the timescale $\mathcal{T} = e/\epsilon$, for which we can derive, on the basis of (8.20), that:

$$\frac{\partial \mathcal{T}}{\partial t} = c_{2\epsilon} - 1, \tag{8.21}$$

with the solution:

$$\mathcal{T} = \mathcal{T}_0 + (c_{2\epsilon} - 1)\, t. \tag{8.22}$$

An obvious condition for *decaying* turbulence is that: $c_{2\epsilon} > 1$. This means that the timescale $\mathcal{T}$ increases linearly as a function of t. We can interpret this as the 'growth' of the turbulent eddies (or rather, the smaller eddies decay first, while the larger eddies remain; because of this, the apparent time scale and length scale of the turbulence increases). This increase of scales is clearly visible in Fig. 4.7. If we substitute (8.22) in (8.20), we get for e:

$$e = (A\, t + B)^{\frac{1}{1-c_{2\epsilon}}}, \tag{8.23}$$

where A and B are determined by the initial conditions. Based on experimental data it follows that the exponent in (8.23) is about equal to -1.09, so that: $c_{2\epsilon} = 1.92$.

The second flow to which the k-ϵ model should comply with is the logarithmic layer in a turbulent channel flow that we encountered in Sect. 6.2. Recall that the logarithmic layer is universal to all wall-bounded turbulent flows. When we apply (8.19) to this flow, we find that:

$$u_*^2 \equiv -\overline{u'v'} = c_\mu \frac{e^2}{\epsilon} \frac{\partial \overline{u}}{\partial y},$$

$$0 = -\overline{u'v'} \frac{\partial \overline{u}}{\partial y} - \epsilon,$$

$$0 = \frac{\partial}{\partial y}\left(\frac{K}{\sigma_\epsilon} \frac{\partial \epsilon}{\partial y}\right) + \frac{\epsilon}{e}(c_{1\epsilon} P - c_{2\epsilon}\epsilon).$$

When we substitute the logarithmic velocity profile (6.15), i.e.:

$$\frac{\overline{u}}{u_*} = \frac{1}{k} \ln\left(\frac{y}{y_0}\right),$$

it follows that:

$$c_\mu = \left(\frac{u_*^2}{e}\right)^2, \quad \text{and:} \quad c_{1\epsilon} = c_{2\epsilon} - \frac{k^2}{\sigma_\epsilon \sqrt{c_\mu}}.$$

Based on experiment data, we find: $c_\mu = 0.09$. Finally, we choose for the constants σ_k and σ_ϵ the values 1 and 1.3, respectively.

With this, all constants in the k-ϵ model For completeness we summarize here the model constants:

$$c_\mu = 0.09, \quad \sigma_k = 1, \quad \sigma_\epsilon = 1.3, \quad c_{2\epsilon} = 1.92, \quad c_{1\epsilon} = 1.44, \quad k = 0.43. \quad (8.24)$$

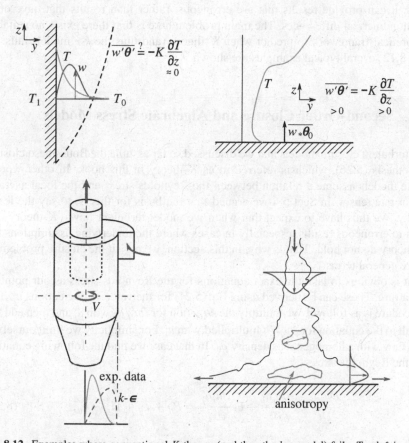

Fig. 8.12 Examples where conventional K-theory (and thus the k-ϵ model) fails. *Top left* in *convection along a vertical wall* the turbulent transport $\overline{w'\theta'}$ is in a direction normal to the direction with the strongest temperature gradient (that is, $\partial T/\partial y$); K-theory would assume that the transport is proportional to $\partial T/\partial z$, which nearly vanishes. *Top right* in *penetrative convection in a stable atmosphere* (with $\partial T/\partial z > 0$) with a positive turbulent heat flux at the surface $(\overline{w_*\theta_0} > 0)$ there is a counter-gradient transport, whereas K-theory is based on gradient transport. *Bottom left* a *cyclone* is used in the process industry to separate droplets and solid particles from a fluid by inducing a strongly swirling flow. Experimental results show that the peak in the tangential velocity component occurs near the centerline of the cyclone, whereas CFD results with the k-ϵ model show a solid body rotation in the central part of the device (Gronald and Derksen 2011). *Bottom right* for an *impinging jet* the flow experiences a very high deformation rate where the turbulence is strongly compressed in the direction normal to the wall and highly stretched in the plane parallel to the wall. This induces a strong anisotropy, which violates the assumption of local equilibrium that underlies the k-ϵ model

The k-ϵ model is applicable to many practical turbulent flow problems. It is the most widespread model found in commercial and open-source codes for computational fluid dynamics, or CFD.

Although, K-theory, and with that the k-ϵ model, provide a suitable approach in many CFD computations, there can be situations where it fails. Often, when it fails, it can provide results that are erroneous, rather than results that have only slight numerical differences. The main problem here is that there exists no suitable theoretical framework to predict when K-theory (and thus the k-ϵ model) fails. In Fig. 8.12 several typical examples are shown.

8.5 Second-Order Closure and Algebraic Stress Models

All turbulent closure models that we discussed so far assume the Boussinesq closure hypothesis (5.26), which is referred to as K-theory in this book. In other words, these models assume a relation between the Reynolds stress and the local average strain rate tensor. In Sect. 5.4 we argued that the basis for this is, to say the least, shaky. We thus have to expect that when we model turbulent flows, K-theory can lead to erroneous results, especially in cases where the underlying assumptions for K-theory do not hold. That is why, in this section, we revisit the closure problem in more general terms.

It is obvious to take the exact equations for the Reynolds stress as our point of departure. These can be derived using Eq. (5.21) for the velocity fluctuations u'_i. The procedure is as follows. We multiply the equation for $\partial u'_i/\partial t$ with u'_j and then add the result to the equation for $\partial u'_j/\partial t$ multiplied with u'_i. For simplicity we limit ourselves to a flow with a homogeneous density ρ_0. In that case we find the following equation for the Reynolds stress:

$$\frac{D\overline{u'_i u'_j}}{Dt} \equiv \frac{\partial \overline{u'_i u'_j}}{\partial t} + \overline{u}_k \frac{\partial \overline{u'_i u'_j}}{\partial x_k} = P_{ij} + T_{ij} + \Pi_{ij} - \epsilon_{ij}, \tag{8.25}$$

with:

$$P_{ij} = -\overline{u'_i u'_k}\frac{\partial \overline{u}_j}{\partial x_k} - \overline{u'_j u'_k}\frac{\partial \overline{u}_i}{\partial x_k},$$

$$T_{ij} = -\frac{\partial}{\partial x_k}\left(\frac{1}{\rho_0}\overline{p' u'_j}\delta_{ik} + \frac{1}{\rho_0}\overline{p' u'_i}\delta_{jk} + \overline{u'_i u'_j u'_k} - \nu\frac{\partial \overline{u'_i u'_j}}{\partial x_k}\right),$$

$$\Pi_{ij} = \frac{1}{\rho_0}\overline{p'\left(\frac{\partial u'_i}{\partial x_j} + \frac{\partial u'_j}{\partial x_i}\right)},$$

$$\epsilon_{ij} = 2\nu\overline{\frac{\partial u'_i}{\partial x_k}\frac{\partial u'_j}{\partial x_k}}.$$

Equation (8.25) describes the spatial and temporal variation of the Reynolds stress due to the physical processes represented by the terms at the right-hand side. Notice that under contraction (that is, for $i = j$) this equation becomes identical to the equation for the turbulent kinetic energy, save for a multiplication factor 2.

It should be clear that by formulating the expression for $\overline{u_i' u_j'}$ we cannot avoid the closure problem. Most of the terms on the right-hand side of (8.25) contain new unknown variables and thus need a closure hypothesis. However, the closure problem has been moved to an equation of a higher order. Instead of closure in the equation for the average velocity (i.e., the *first* statistical moment of the velocity) we now have to formulate a closure for the terms in the equation for the *second* moment. That is why we refer to this as a *second-order closure hypothesis*. The motivation for this is of course the assumption (or expectation) that our flow problem would be less sensitive to a higher-order closure hypothesis than to a first-order closure. (We could interpret this as a sort of asymptotic expansion: approximations for the higher-order terms are considered to be less important than those of lower order.) However, a proof for this assumption is absent.

We now discuss the various terms in (8.25) and formulate a closure hypothesis when needed.

P_{ij}: **production** This term describes the production of Reynolds stress by gradients in the average velocity field. We notice that for this term no closure hypothesis is needed; in other words, the production of Reynolds stress is described exactly by the second-order closure model. This fact is often mentioned as the most important reason to use second-order closure.

ϵ_{ij}: **molecular destruction** This term describes the destruction of Reynolds stress by molecular viscosity. On the basis of the discussion from Chap. 7 we know that these molecular effects take place at the microstructure. Also, we saw in Sect. 8.3 that for large Reynolds numbers, the microstructure can be considered be in local equilibrium, and therefore we concluded that the microstructure is isotropic. This implies that ϵ_{ij} has to be an isotropic tensor. When we use the fact that (8.25) turns into the equation for the turbulent kinetic energy for $i = j$, it follows that:

$$\epsilon_{ij} = \frac{2}{3} \epsilon \, \delta_{ij}, \tag{8.26}$$

where ϵ represents the viscous dissipation of turbulent kinetic energy. We have already encountered in Sect. 7.4 a comparable result for the energy equation per component (7.19). This implies for the Reynolds stress that the viscous destruction of this stress equals zero, with the argument that at the microscales, where viscous destruction takes place, the Reynolds stress due to the condition of local isotropy is negligible.

T_{ij}: **transport** As mentioned several times before, this divergence term fulfills the function of a spatial distribution of, in this case, the Reynolds stress. We see that this term consists of several contributions, namely: transport by velocity fluctuations and transport by pressure fluctuations. Transport by molecular effects is negligible here, based on the same arguments given for the transport term in the kinetic energy Eq. (7.8).

There are several closure hypotheses for the transport term, and almost all of them are based on the gradient hypothesis. For transport by velocity fluctuations, the closure hypothesis that is mostly applied reads:

$$\overline{u_i' u_j' u_k'} = -c_s \frac{e}{\epsilon} \, \overline{u_k' u_m'} \, \frac{\partial \overline{u_i' u_j'}}{\partial x_m}, \tag{8.27}$$

where c_s is an empirical constant, for which the value of 0.22 is chosen. In this equation, the term e/ϵ represents again a characteristic timescale. A closure hypothesis such as (8.27) can in principle be avoided by deducing an equation for $\overline{u_i' u_j' u_k'}$. In this way we reach a so-called *third-order closure model*. However, such an expansion is not often applied, because the third-order equations, and the subsequent closure hypotheses in these equations, can become very complicated without really leading to any substantial improvement for the model results.

Often, no separate closure hypothesis is formulated for transport by pressure fluctuations. In other words, the effect of pressure fluctuations is incorporated in the closure hypothesis (8.27). Transport by molecular diffusion is often neglected.

Π_{ij}: **pressure-velocity correlation** For this term we can formally derive an expression, but for this we need to assume an expression for the pressure fluctuations p'. We can find this expression by taking the divergence of Eq. (5.21) for u_i' and using the fact that the velocity field is incompressible, i.e. $\partial u_i'/\partial x_i = 0$. This results in a so-called *Poisson equation* for p', which reads

$$\frac{1}{\rho_0} \frac{\partial^2 p'}{\partial x_m^2} = -2 \frac{\partial \overline{u}_k}{\partial x_l} \frac{\partial u_l'}{\partial x_k} - \frac{\partial u_l'}{\partial x_k} \frac{\partial u_k'}{\partial x_l} - \frac{\partial \overline{u_l' u_k'}}{\partial x_l \partial x_k}. \tag{8.28}$$

For the Poisson equation:

$$\frac{1}{\rho_0} \frac{\partial^2 p'}{\partial x_m^2} = f(\underline{x}),$$

the following exact solution exists:

$$\frac{1}{\rho_0} p'(\underline{x}) = -\frac{1}{4\pi} \iiint\limits_V \frac{f(\underline{x}')}{|\underline{x} - \underline{x}'|} \, d\underline{x}',$$

where the integral is taken over the entire volume V that encompasses the flow. Here we implicitly assumed that: $f(\underline{x}) = 0$ at the boundaries of V, or: $f(\underline{x}) \to 0$ for $|\underline{x}| \to \infty$. Using this solution we can obtain an expression for Π_{ij} in terms of volume integrals over various combinations of the velocity fluctuations. It follows that the term Π_{ij} can be constructed from two contributions: $\Pi_{ij} = \Pi_{ij}^{(1)} + \Pi_{ij}^{(2)}$:

$$\Pi_{ij}^{(1)}(\underline{x}) = \frac{1}{4\pi} \iiint\limits_V \overline{\frac{\partial u_k'}{\partial x_l'} \frac{\partial u_l'}{\partial x_k'} \left(\frac{\partial u_i'}{\partial x_j} + \frac{\partial u_j'}{\partial x_i} \right)} \frac{dx'}{|\underline{x} - \underline{x}'|},$$

$$\Pi_{ij}^{(2)}(\underline{x}) = \frac{1}{4\pi} \iiint\limits_V 2\frac{\partial \overline{u}_k}{\partial x_l'} \overline{\frac{\partial u_l'}{\partial x_k'} \left(\frac{\partial u_i'}{\partial x_j} + \frac{\partial u_j'}{\partial x_i} \right)} \frac{dx'}{|\underline{x} - \underline{x}'|}.$$

We have already encountered Π_{ij} in Sect. 7.4 when we discussed the energy equation per component. We showed there that this term 'restores isotropy.' This result, better known as *Rotta's hypothesis*, here forms the basis of the closure hypothesis for $\Pi_{ij}^{(1)}$, with the result:

$$\Pi_{ij}^{(1)} = -c_1 \frac{\epsilon}{e} \left(\overline{u_i' u_j'} - \frac{2}{3} \delta_{ij} e \right). \tag{8.29}$$

The term $\Pi_{ij}^{(2)}$ describes that the pressure-velocity correlation also depends on the gradients of the average velocity. The first step that is often taken to a find solution of this term is to assume that the term $\partial \overline{u}_k / \partial x_l'$ is constant, so that it can be excluded from the integration. The remainder of the integral can then be evaluated using correlation functions (see Chap. 9). A complete calculation can be pretty complicated, and therefore the closure hypothesis is often simplified to:

$$\Pi_{ij}^{(2)} = -c_2 \left\{ \left(P_{ij} - \frac{1}{3} \delta_{ij} P_{kk} \right) - \left(C_{ij} - \frac{1}{3} \delta_{ij} C_{kk} \right) \right\}, \tag{8.30}$$

where P_{ij} represents the production term in (8.25) and C_{ij} the advection term:

$$C_{ij} = \overline{u}_k \frac{\partial \overline{u_i' u_j'}}{\partial x_k}.$$

The effect of this first term in (8.30) is that the production of Reynolds stress is decreased. In other words, the pressure-velocity correlation suppresses the effectiveness of the Reynolds-stress production.

For the constants c_1 and c_2 in (8.29) and (8.30), respectively, various values have been proposed. This is illustrated in Fig. 8.13, where c_1 is plotted against c_2. Note that these all appear to lie on a straight line (see Problem 1 at the end of this section). It is recommended to use: $c_1 = 1.8$, and: $c_2 = 0.6$.

With this, the equation for Reynolds stress is closed, and our second-order closure model is complete. As an illustration, we apply this model to the channel flow, to which we already have become familiar with, and for which we have: $\overline{u}_i = (\overline{u}_1(x_2), 0, 0)$. We use the approximations: $D\overline{u_i' u_j'}/Dt \approx 0$, and $T_{ij} \approx 0$. These approximations appear to hold quite well in the region close to the wall. The equation for the Reynolds stress component $\overline{u_1' u_2'}$ then reduces to

Fig. 8.13 Closure constants in the pressure-velocity correlation

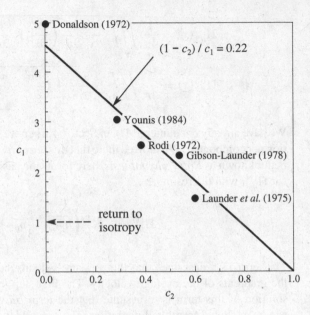

$$0 = -(1 - c_2)\overline{u_2'^2}\frac{\partial \overline{u}_1}{\partial x_2} - c_1\frac{\epsilon}{e}\overline{u_1'u_2'},$$

with the solution:

$$-\overline{u_1'u_2'} = \frac{(1 - c_2)}{c_1}\frac{e}{\epsilon}\overline{u_2'^2}\frac{\partial \overline{u}_1}{\partial x_2}.$$

We essentially retrieve the K-theory result. In other words, the second-order closure model yields K-theory as the limit solution. This result also suggests the circumstances for which K-theory would no longer hold, namely: when $D\overline{u_i'u_j'}/Dt$ and T_{ij} are no longer negligible. This is the case when turbulence is strongly non-local in time and/or in space. This limitation is not surprising given our discussion on the validity of K-theory in Sect. 5.4. For those cases we thus have to solve the complete second-order closure model.

However, we have to realize that a solution of the complete second-order closure model is significantly more complicated. Apart from the three equations for the average velocity $\overline{u}_i$, we also need to solve six equations for the components of the Reynolds stress tensor $\overline{u_i'u_j'}$, and an additional equation for ϵ. Thus, we are confronted now with a system of 10 coupled, nonlinear partial differential equations. Apart from an substantial computational effort, we are also confronted with many numerical complexities.

Thus, a simplification of the second-order closure model is highly desirable. The most common simplification is the so-called *algebraic stress model*. As its name clearly suggests, the partial differential equations for the Reynolds stresses are reduced to more simple algebraic equations. The basis of this model is formed by the following approximation:

$$\frac{D\overline{u_i'u_j'}}{Dt} - T_{ij} = \frac{\overline{u_i'u_j'}}{e}\left(\frac{De}{Dt} - T_k\right), \tag{8.31}$$

where T_{ij} is the transport term in the Reynolds stress equation, and T_k the transport term in Eq. (7.8) for the turbulent kinetic energy. The advection and transport in the Reynolds stress equation are taken to be proportional to the advection and transport in the kinetic energy equation. The proportionality constant is then related to the anisotropy in the Reynolds stress. Using the equation for e, it then follows that:

$$\frac{D\overline{u_i'u_j'}}{Dt} - T_{ij} = \frac{\overline{u_i'u_j'}}{e}\left(P_k - \epsilon\right), \tag{8.32}$$

where $P_k = \frac{1}{2}P_{ii}$ is the production of turbulent kinetic energy. After substituting (8.32) in (8.25), using the closure hypotheses (8.26) for ϵ_{ij}, and after substituting (8.29) and (8.30) for $\Pi_{ij}^{(1)}$ and $\Pi_{ij}^{(2)}$, respectively, we arrive at the following implicit equation for the Reynolds stress:

$$\overline{u_i'u_j'} - \frac{2}{3}e\,\delta_{ij} = \frac{(1-c_2)(P_{ij} - \frac{2}{3}\delta_{ij}P_{kk})}{P_k - \epsilon + c_1\epsilon}. \tag{8.33}$$

Actual computations demonstrate that the algebraic stress model in wall-bounded turbulence yields results that can be compared with results for the complete second-order closure model. However, for free turbulence, in particular in the case of an axisymmetric flow geometry, we find large differences. In those situations we need to solve the problem through the full second-order closure model.

Problems

1. Apply the equation for $\overline{u_i'u_j'}$ and the closure hypotheses discussed above to the logarithmic layer in a turbulent channel flow. It can be assumed that: $D\overline{u_i'u_j'}/Dt \approx 0$, and: $T_{ij} \approx 0$. Show that we then have the following relation for the constants c_1 and c_2:

$$\frac{1-c_2}{c_1} = \frac{u_*^4}{e\,\overline{u_2'}^2}.$$

2. Derive Eq. (8.33).
3. Consider decaying grid turbulence. Apply the equations for $\overline{u_i'u_j'}$ and the closure hypotheses for Π_{ij} to the turbulent kinetic energy for each velocity component, that is:

$$\overline{u_1'}^2, \ \overline{u_2'}^2 \ \text{and:} \ \overline{u_3'}^2.$$

At time $t = 0$, we have:

$$\overline{u_1'}^2 \neq \overline{u_2'}^2 \neq \overline{u_3'}^2.$$

Show that this flow, for $t \to \infty$, can only reach an isotropic state, that is:

$$\overline{u_1'}^2 = \overline{u_2'}^2 = \overline{u_3'}^2,$$

when $c_1 > 1$.

8.6 Large Eddy Simulation of Turbulence

In Sect. 4.3 we discussed the method of *direct numerical simulation* (DNS) to study turbulent flows. This is a very powerful numerical technique that resolves all scales of motion in a turbulent flow. However, even with today's fast (super-) computing systems, direct numerical simulation is still limited to low Reynolds number flows. However, in many cases it is not necessary to capture all scales of motion in a turbulent flow. If we are, for instance, interested in the lift and drag forces on an object, we only have to consider the large scales, because these scales carry most of the momentum. In a *large-eddy simulation* (LES) we only resolve the large scales in the flow, which reduces the computational costs considerably.

The first step in a *large-eddy simulation* is to eliminate the small scales (microstructure) from the problem. This is possible by filtering the turbulent field. Formally, we can write this filter operation as:

$$[f(x_1, x_2, x_3)] = \iiint\limits_V G(\underline{\xi} - \underline{x}) f(\underline{x}) \, d\xi_1 \, d\xi_2 \, d\xi_3, \qquad (8.34)$$

where $G(\underline{\xi} - \underline{x})$ represents the *filter function*. A commonly used filter is the 'top hat' filter. This filter has a value of $1/V_f$ inside the volume $V_f = \Delta_f^3$, while it is zero outside this volume, i.e.:

$$G(\underline{\xi} - \underline{x}) = \begin{cases} V_f^{-1} & \text{for:} \quad \underline{\xi} - \underline{x} \in V_f, \\ 0, & \text{elsewhere,} \end{cases}$$

where the integration volume V_f is centered around the point (x_1, x_2, x_3). The term Δ_f is usually called the *filter length*. By this filter operation, for which we use the notation: $[\cdots]$, all fluctuations with a scale *smaller* than the filter length Δ_f are removed. When we adjust Δ_f to the *characteristic size* of our *numerical grid* Δ, it follows that the filtered velocity field represents the macrostructure that we can represent with our numerical grid. For this reason the filtered variables are sometimes referred to as the 'resolved' scales.

We have to formulate equations for the filtered variables that should be solved numerically. We find these equations by applying the filter operation in (8.34) to the Navier–Stokes equations. Before we perform this operation, we first rewrite the non-linear term and the viscous term as:

$$u_j \frac{\partial u_i}{\partial x_j} = \frac{\partial u_i u_j}{\partial x_j}, \quad \text{and:} \quad \nu \frac{\partial^2 u_i}{\partial x_j^2} = \frac{\partial}{\partial x_j} \nu \left(\frac{\partial u_i}{\partial x_j} + \frac{\partial u_j}{\partial x_i} \right),$$

respectively, where we have used the fact that the flow is *incompressible*, that is: $\partial u_i / \partial x_i = 0$. The *filtered* Navier–Stokes equations can now be written as:

$$\frac{\partial [u_i]}{\partial t} + \frac{\partial [u_i u_j]}{\partial x_j} = -\frac{\partial [p]}{\partial x_i} + \frac{\partial}{\partial x_j} \nu \left(\frac{\partial [u_i]}{\partial x_j} + \frac{\partial [u_j]}{\partial x_i} \right), \qquad (8.35)$$

where the filtering operation has introduced a new unknown quantity: $[u_i u_j]$. This new unknown quantity can be written as:

$$[u_i u_j] = [u_i][u_j] + \underbrace{[u_i u_j] - [u_i][u_j]}_{\text{subgrid stress}}.$$

For the last term on the right hand side we introduce a new symbol, the so-called *subgrid stress*:

$$\tau_{\text{sgs}} = -\rho_0 \left([u_i u_j] - [u_i][u_j] \right) \qquad (8.36)$$

They idea behind large eddy simulation is that the difference between $[u_i u_j]$ and $[u_i][u_j]$, i.e. the *subgrid stress* τ_{sgs}, is small because of the limited extend of the spatial filter. The *filtered* Navier–Stokes equations now read:

$$\frac{\partial [u_i]}{\partial t} + \frac{\partial [u_i][u_j]}{\partial x_j} = -\frac{1}{\rho_0} \frac{\partial [p]}{\partial x_i} + \frac{\partial}{\partial x_j} \nu \left(\frac{\partial [u_i]}{\partial x_j} + \frac{\partial [u_j]}{\partial x_i} \right) + \frac{1}{\rho_0} \frac{\partial \tau_{\text{sgs}}}{\partial x_j}, \qquad (8.37)$$

The subgrid stress in (8.36) expresses how the filtered microstructure exerts an effective stress to the large eddies. In principle, this term can be compared to the Reynolds stress $-\overline{u_i' u_j'}$. However, here τ_{sgs} only describes the stress of the *microstructure*, while $-\overline{u_i' u_j'}$ is the stress on the average flow due to *all* the turbulence scales.

Again, we are confronted with a *closure problem*, because in order to solve Eq. (8.37) we have to specify a closure relation for (8.36). In most cases the *Prandtl mixing length hypothesis*, or its generalized form: the *Smagorinsky model* (7.18), can be used. Let us consider here the Smagorinsky model for τ_{sgs}:

$$\tau_{\text{sgs}} = \tau_{ij}^{(s)} = K_s \left(\frac{\partial [u_i]}{\partial x_j} + \frac{\partial [u_j]}{\partial x_i} \right),$$

where K_s is the *subgrid eddy viscosity*. We can rewrite the last two terms on the right-hand side of (8.37) as:

$$\frac{\partial}{\partial x_j} \nu \left(\frac{\partial [u_i]}{\partial x_j} + \frac{\partial [u_j]}{\partial x_i} \right) + \frac{\partial \tau_{\text{sgs}}}{\partial x_j} = \frac{\partial}{\partial x_j} (\nu + K_s) \left(\frac{\partial [u_i]}{\partial x_j} + \frac{\partial [u_j]}{\partial x_i} \right).$$

For the eddy viscosity K_s the following relation can be used:

$$K_s = \mathcal{L}^2 \left| \frac{\partial [u_i]}{\partial x_j} + \frac{\partial [u_j]}{\partial x_i} \right|. \tag{8.38}$$

Based on the kinetic energy equation it can be demonstrated that the *mixing length* $\mathcal{L}$ is in this case proportional to the filter length Δ_f, that is: $\mathcal{L} = \beta \, \Delta_f$, where the proportionality constant $\beta = \mathcal{O}(1)$ depends on the type of filter. We now introduce the constant C_s, which is defined as:

$$C_s = \frac{\mathcal{L}}{\Delta}, \tag{8.39}$$

where Δ is the *characteristic grid distance*. Using C_s we can control the *effective resolution* of the large eddy simulation. For small values of C_s, $\mathcal{L}$ (and thus also Δ_f) is small compared to the numerical grid. This means that the grid is too coarse to solve all fluctuations; we can expect some influence of numerical truncation errors. On the other hand, when C_s is large, it follows that $\mathcal{L}$ (and thus also Δ_f) is large compared to the numerical grid. The grid is then sufficient to represent all large eddies. You could even say that when we increase C_s too much, we filter away too many fluctuations. The optimal state is found when $\Delta_f \sim \Delta$. On the basis of a theoretical consideration it then follows that $C_s \sim 0.1$–0.2 (see Problem 1 at the end of this section).

Although we computed a few things, it seems that we did not make much progress using our large eddy model as compared to the turbulence models that we discussed before. Indeed, we still have to use a *closure hypothesis*. However, there is an important difference. So far we have discussed the closure of *average quantities*, such as the Reynolds stress, which is mainly determined by the macrostructure. In a large eddy simulation we compute the macrostructure explicitly, and we only have to specify closure for the effects of the subgrid scales, or the *microstructure*. This is much simpler, because in the next chapter we shall see that the microscales have a *universal* structure, which can be described with a relatively simple theory. Apart from this it appears from simulation results using large eddy models that the statistics for large simulated structures are relatively insensitive to the subgrid closure model. This means that, despite possible errors in the closure hypothesis of the microstructure, large eddies can be computed reliably, which is, after all, the objective of the large eddy model.

The principal advantage of a large eddy model is thus that we can simulate flows at large Reynolds numbers with relatively modest computer capacity. We illustrate his by presenting some of the simulation results for a plane channel flow with a Reynolds number of 13,800. In Fig. 8.14 are presented some of the results in the form of a contour plot of the velocity fluctuations in the flow direction. In this graph some clearly elongated structures can be recognized, which we can associate with the *low-speed streaks* discussed in Sect. 8.2. Another result obtained by means of large eddy simulation is shown in Fig. 8.15. In this case small particles parallel to the wall are released in the simulated flow domain. This visualization of the simulation

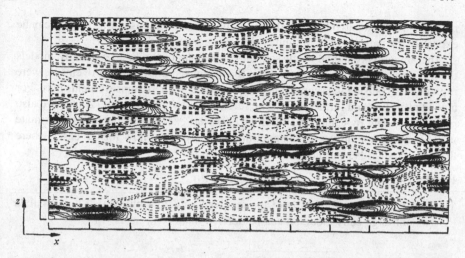

Fig. 8.14 Low-speed streaks in a large eddy simulation of a channel flow at a Reynolds number of 18,300 (based on the centerline velocity and channel half-width H). The contours indicate the fluctuations of the streamwise component of the velocity in a plane at a distance of $y^+ = 6.26$ parallel to the channel wall; the uninterrupted contour lines indicate regions with a flow speed that is less than the local mean velocity. The streamwise extend of the figure is $2\pi H(4021\nu/u_\tau)$, and its spanwise extent is πH. From: Moin and Kim (1982)

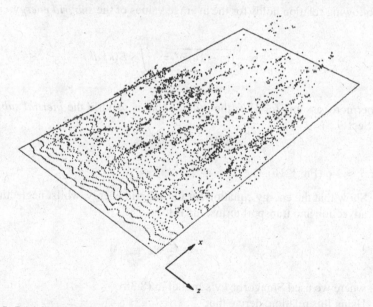

Fig. 8.15 Particle trajectories in wall turbulence, simulated using the large eddy model described in Fig. 8.14. The particles are generated from a 'z-wire' located at $y^+ = 12$ (cf. Fig. 8.5). From: Moin and Kim (1982)

results can be compared to the hydrogen bubble visualizations in Fig. 8.5. As can be expected, the patterns of both figures show a clear resemblance.

The Smagorisky model discussed above was one of the first large eddy models that was described in the literature. Over the recent years several improvements were reported. It is outside the scope of this book to discuss these improvements here. For a good overview and further details we refer to the book by Sagaut (2005). We also refer to a review by Cabot and Moin (1999) who examines different approximate wall boundary conditions for channel flow and separated flow for the cases where the numerical grids near the wall are not resolved.

Problem

1. For the average deformation of the resolved velocity field, which we calculate with our large eddy model, it can be derived that:

$$\overline{\left(\frac{\partial [u_i]}{\partial x_i} + \frac{\partial [u_j]}{\partial x_j}\right)^2} = S^2 \approx 4 \int_0^{2\pi/\Delta_f} k^2 E(k)\, dk,$$

where the *energy spectrum* $E(k)$ is defined in (9.19). This relation is exact in the case of *isotropic turbulence*, which we discuss in Sect. 9.6. It can be shown that the following relation holds for the average values of the *subgrid energy*:

$$\overline{e^{(s)}} \equiv \frac{1}{2}\overline{(u_i - [u_i])^2} = \int_{2\pi/\Delta_f}^{\infty} E(k)\, dk.$$

Furthermore, assume for $E(k)$ the relation that is valid in the *inertial subrange* (see Sect. 9.5):

$$E(k) = \beta \epsilon^{2/3} k^{-5/3},$$

with $\beta \approx 1.6$ (Fig. 8.16).

(a) Show that the energy equation for the subgrid energy, whilst neglecting the advection and transport terms, leads to:

$$\epsilon = \frac{1}{2} S^3 \ell^2,$$

where we used Smagorinsky's model in (8.38).

(b) Using this relation, derive that:

$$\frac{\mathcal{L}}{\Delta_f} = 0.17,$$

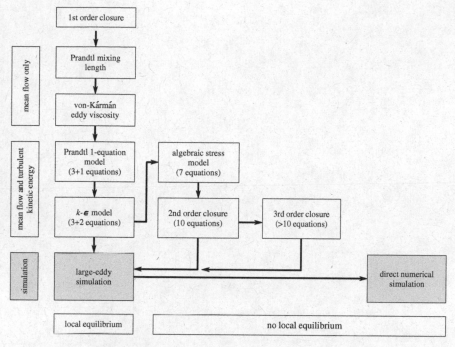

Fig. 8.16 Summary of turbulence closure models. *White boxes* refer to *Reynolds-averaged Navier–Stokes* (RANS) models; *shaded boxes* refer to simulation methods, such as *large-eddy simulation* (LES) and *direct numerical simulation* (DNS). RANS approaches where the flow (ensemble) statistics, such as the mean velocity and turbulent kinetic energy, can vary as a function of time are referred to as *unsteady* RANS, or URANS. The *arrows* indicate an increasing computational effort

from which it follows that:

$$C_s = 0.17 \frac{\Delta_f}{\Delta}.$$

(c) Show that:

$$\epsilon = 1.69 \frac{\overline{e^{(s)}}^{3/2}}{\Delta_f}.$$

Chapter 9
Correlation Function and Spectrum

So far we have mainly dealt with the so-called *single-point statistical moments* of the turbulent flow. These are statistical quantities, such as the average or the variance, defined at a single point in space (see Chap. 5). However, these single-point moments do not describe the spatial and temporal *structure* of turbulence. For this we need *multiple-point moments* that describe the relation between the various variables in space and time.

The simplest multiple-point moment is the *correlation function* of two variables at two different points. We formally defined such a correlation in (5.12). In this chapter we focus on this correlation function and its *Fourier transform*, referred to as the *spectrum*.

9.1 Time Correlations

Consider a turbulent variable as a function of time. Take for example the fluctuating velocity component $u'(t)$ with $\overline{u'} = 0$. There are two ways to define such a variable as a function of time:

Eulerian: Here we measure u' at a fixed point as a function of t. You could think for example of an instrument that is permanently installed in a flow (see Sect. 4.3), and with which we register turbulent fluctuations as a function of time.

Lagrangian: In this case we do not have a fixed coordinate system, but we move along with a material fluid element on its way through the flow. As it moves along we register the turbulent velocity fluctuations as a function of time.

The latter description has some theoretical advantages for certain applications. The measurement of the Lagrangian velocity fluctuations can in practice only be done through *particle tracking velocimetry*. Later we return to the properties of the Eulerian and Lagrangian measurements. For now it only matters that we define the turbulent variable $u'(t)$, which we interpret as a velocity fluctuation as a function of time.

© Springer International Publishing Switzerland 2016   183
F.T.M. Nieuwstadt et al., *Turbulence*, DOI 10.1007/978-3-319-31599-7_9

When we consider $u'(t)$ at times t_1 and t_2, we can introduce the following time correlation:

$$\overline{u'(t_1)u'(t_2)}. \tag{9.1}$$

Because we consider the same velocity components at the two times we call this the *autocorrelation function*. For a formal definition of this correlation in terms of the probability distribution we refer to Sect. 5.1. Here we limit ourselves to a *stationary process*, or, *stationary turbulence*. For this we found in Chap. 5 that the correlation is only a function of the *difference* in time, or:

$$\overline{u'(t_1)u'(t_2)} = R(t_2 - t_1) = R(\tau) = \overline{u'^2}\rho(\tau), \tag{9.2}$$

where: $\tau = t_2 - t_1$. The term $\rho(\tau)$ is called the *correlation coefficient*. For a stationary process, $\overline{u'^2}$ is, by definition, *independent* of t. (It should be clear that we can define a correlation like (9.1) for any other time-varying scalar variable in a turbulent flow.) Some properties of $\rho(\tau)$ are:

$$
\begin{aligned}
&i : \rho(0) = 1, \\
&ii : |\rho(\tau)| \leqslant \rho(0) = 1 \quad \forall\, t, \\
&iii : \rho(\tau) = \rho(-\tau), \\
&iv : \rho(\tau) \to 0 \quad \text{for:} \quad \tau \to \infty.
\end{aligned}
\tag{9.3}
$$

The last property means that the correlation vanishes for large time differences. In other words, with a given velocity, it is impossible to predict the velocity for a large time difference. We saw in Chap. 1 that this is an essential property of chaos. In short, a chaotic process, and thus turbulence, has a *finite* timescale. We can quantify this timescale, but only when $\rho(\tau)$ approaches zero fast enough when $\tau \to \infty$, using:

$$\mathcal{T} = \int_0^\infty \rho(\tau)\, d\tau, \tag{9.4}$$

where $\mathcal{T}$ is called the *integral timescale*. This timescale is a measure for the time difference over which the significant correlation persists. It is therefore obvious to identify the $\mathcal{T}$ with the timescale of the *macrostructure*.

Next, we consider the correlation function of the time derivatives:

$$\overline{\frac{\partial u'(t_1)}{\partial t_1}\frac{\partial u'(t_2)}{\partial t_2}}. \tag{9.5}$$

This correlation function can be related to the correlation function for the velocities that we introduced above. For a *stationary process* it follows on the basis of $\tau = t_2 - t_1$ that

$$\frac{\partial}{\partial t_1} = -\frac{\partial}{\partial \tau}, \quad \text{and:} \quad \frac{\partial}{\partial t_2} = \frac{\partial}{\partial \tau},$$

so that:

$$\overline{\frac{\partial u'(t_1)}{\partial t_1}\frac{\partial u'(t_2)}{\partial t_2}} = -\overline{u'^2}\frac{\partial^2 \rho(\tau)}{\partial \tau^2}. \tag{9.6}$$

When: $t_1 = t_2$, (or: $\tau = 0$) we now find that:

$$\overline{\left(\frac{\partial u'}{\partial t}\right)^2} = -\overline{u'^2}\left.\frac{\partial^2 \rho(\tau)}{\partial \tau^2}\right|_{\tau=0}. \tag{9.7}$$

In Chap. 7 we argued that the differential of a fluctuation, as it occurs on the left-hand side of (9.7), is dominated by the *microstructure*, and in the same chapter we introduced the following scaling:

$$\overline{\left(\frac{\partial u'}{\partial t}\right)^2} = \frac{\overline{u'^2}}{\lambda_T^2}, \tag{9.8}$$

where λ_T is the *Taylor micro-timescale*.[1] Based on (9.7), we find that this Taylor microscale is coupled to the second derivative of $\rho(\tau)$ at $\tau = 0$. We can interpret this second derivative as the *curvature* of the correlation function at $\tau = 0$, since we can write, based on a power series expansion, that:

$$\rho(\tau) = \rho(0) + \frac{1}{2}\tau^2\left.\frac{\partial^2 \rho}{\partial \tau^2}\right|_{\tau=0} + \cdots, \tag{9.9}$$

where, based on property *iii* in (9.3), the term with $\partial \rho/\partial \tau$ equals zero. In other words, the behavior of the correlation function for $\tau \approx 0$ relates to the *microstructure*.

With this we demonstrated that the autocorrelation function (or the autocorrelation coefficient function) contains information about the temporal structure of turbulence. For $\tau \sim \mathcal{T}$ this correlation describes the macrostructure and for $\tau \sim 0$ the microstructure. This is schematically represented in Fig. 9.1.

Problem

1. Consider the quantity $\overline{\{u'(t_2) - u'(t_1)\}^2} \geqslant 0$, and use it to prove property *ii* in (9.3).

---

[1] Please note that λ_T has the dimension of time and is the Taylor microscale for temporal fluctuations.

Fig. 9.1 The temporal
autocorrelation function of a
turbulent signal

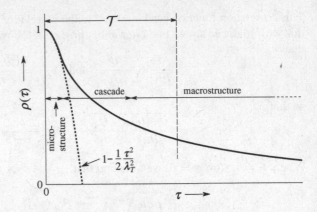

9.2 The Spectrum

Because $\rho(t)$ satisfies property *iv* in (9.3), we can formally introduce the following
Fourier transform pair:

$$S(\omega) = \frac{1}{2\pi} \int\limits_{-\infty}^{\infty} e^{-i\omega\tau} R(\tau)\, d\tau$$

$$R(\tau) = \int\limits_{-\infty}^{\infty} e^{i\omega\tau} S(\omega)\, d\omega. \tag{9.10}$$

The background for this transformation is that, on the basis of the harmonic analysis
of time series, every stationary time series can be represented as a superposition of
Fourier components, that is waves with a period $T = 2\pi/\omega$. The term $S(\omega)$ can then
be interpreted as the square of the amplitude of the wave with angular frequency ω.
$S(\omega)$ is often called the *energy spectrum*, because on the basis of (9.10) for $\tau = 0$
and $R(0) = \overline{u'^2}$ it follows that:

$$\overline{u'^2} = \int\limits_{-\infty}^{\infty} S(\omega)\, d\omega, \tag{9.11}$$

So, $S(\omega)$ contributes to $\overline{u'^2}$, which is here the x-component of the energy, for frequencies between ω and $\omega + d\omega$. When the integral (9.4) exists, or in other words when the
turbulence has a *finite* timescale, it can be shown that the spectrum is *continuous*. If
a process would consist of a finite number of (quasi-)periodic orbits in phase space,
then its spectrum would contain discrete peaks, each associated with a periodic orbit;

hence, a *continuous* spectrum is sometimes referred to as a characteristic property of a *chaotic* process (see also Fig. 3.24).

The spectrum $S(\omega)$ has the following properties:

$$i : S(\omega) > 0 \quad \forall \omega,$$
$$ii : S(\omega) = S(-\omega), \tag{9.12}$$
$$iii : S(\omega) = S^*(\omega),$$

where the superscript $*$ indicates the *complex conjugate*. (On the basis of (9.10), $S(\omega)$ can have a complex value.) When $S(\omega)$ does not satisfy these conditions, it can not be the spectrum of a *stationary* process.

The first property in (9.12) follows from the fact that $S(\omega)$ is equal to the energy of a wave with a frequency ω. The two other properties follow from (9.3) and (9.10), and these properties lead to the conclusion that $S(\omega)$ is an even and real function of ω. Based on this we can simplify the transformation (9.10) to a *cosine transform*, defined as:

$$E(\omega) = \frac{2}{\pi} \int\limits_0^\infty R(\tau) \cos(\omega\tau)\,d\tau,$$

$$R(\tau) = \int\limits_0^\infty E(\omega) \cos(\omega\tau)\,d\omega, \tag{9.13}$$

with: $E(\omega) = 2S(\omega)$. A final property is found by substitution of $\omega = 0$ in (9.10), that is:

$$S(0) = \frac{1}{2\pi} \int\limits_{-\infty}^\infty R(\tau)\,d\tau = \frac{\overline{u'^2}\,\mathcal{T}}{\pi}. \tag{9.14}$$

We defined $\mathcal{T}$ in (9.4) as the *integral timescale* that relates to the macrostructure. This suggests that the behavior of $S(\omega)$ close to $\omega \sim 0$ is representative for the macrostructure. This is in accord with the fact that $\omega \to 0$ corresponds to long waves of which an interpretation in terms of *large eddies* is obvious. However, we need to emphasize that a *wave* and an *eddy* are substantially different; in principle, a wave has an *infinite* extent, while an eddy has definitely *finite* dimensions. That is why we should actually interpret the spectrum $S(\omega)$ for turbulence as *wave packets* of bandwidth $\Delta\omega$ centered around the frequency ω. For a further elaboration on this interpretation we refer to the book of Tennekes and Lumley (1972).

Based on similar arguments, where smaller eddies are associated with smaller time periods (or high frequencies), we can couple the region $\omega \to \infty$ to the *microstructure*. These results for the spectrum are summarized schematically in Fig. 9.2. We have seen that both the correlation $R(\tau)$ and the spectrum $S(\omega)$ contain information about the structure of turbulence. In practice however, we often consider the spectrum, because we can make a more direct link between the eddy size and the frequency.

Fig. 9.2 The turbulent time spectrum with a bandwidth ω_B

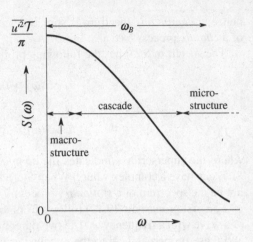

Problems

1. Define the *bandwidth* ω_B of the spectrum as: $\omega_B = (\int_0^\infty S(\omega)\,d\omega)/S(0)$. Find a relation between ω_B and T. Interpret the limits $\omega_B \to 0$ and $\omega_B \to \infty$, both in terms of the spectrum and in terms of the correlation function.
2. Determine the spectra for the following correlation functions:

   (a) $\rho(\tau) = 1$ for $\tau < T$ and $\rho(\tau) = 0$ for $\tau > T$;
   (b) $\rho(\tau) = \exp(-\tau/T)$.

   Which of the spectra is physically realistic? Does the correlation function behave realistically for (b) when $\tau \to 0$?

9.3 Spatial Correlations and Spectra

Now consider the simultaneous measurement of the velocity fluctuations in two points $\underline{x}_1$ and $\underline{x}_2$. With this we can define the *two-point correlation* $\overline{u'_i(\underline{x}_1)u'_j(\underline{x}_2)}$. For *homogeneous turbulence* this quantity is a function of the separation vector $\underline{r} = \underline{x}_2 - \underline{x}_1$ only, so it follows that:

$$R_{ij}(\underline{r}) = \overline{u'_i(\underline{x}_1)u'_j(\underline{x}_2)}, \tag{9.15}$$

which we call the *correlation tensor*. The correlation R_{ij} has the following properties:

$$i : R_{ij}(0) = \overline{u'_i u'_j},$$
$$ii : R_{ij}(\underline{r}) = R_{ji}(-\underline{r}), \tag{9.16}$$
$$iii : R_{ij}(\underline{r}) \to 0 \quad \text{for:} \ \ |\underline{r}| \to \infty,$$

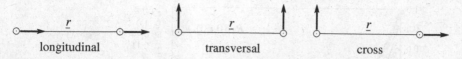

Fig. 9.3 Correlations between the velocities measured at two points

where the last property relates to the fact that the spatial dimensions of a turbulent eddy are finite.

The correlation tensor has in principle nine components of which six are independent. Every one of those components is a function of the separation vector $\underline{r}$. For the tensor components we distinguish between different configurations of the directions of the velocity vectors u'_i and u'_j with respect to the separation vector. These configurations, illustrated in Fig. 9.3, are:

longitudinal correlation: this describes the correlation between two velocity components that are parallel to the separation vector $\underline{r}$ in both points. In case the separation vector is chosen along the x-axis, i.e. $\underline{r} = (r, 0, 0)$, $R_{11}(r) = \overline{u'(x)u'(x+r)}$ is a *longitudinal correlation*; in case $\underline{r} = (0, r, 0)$ lies along the y-axis, it follows that $R_{22}(r)$ is also a longitudinal correlation.

transversal correlation: in this case we consider the correlation between two velocities that have the same direction and are perpendicular to the separation vector $\underline{r}$. In principle there are two transversal correlations. As an example we take again $\underline{r} = (r, 0, 0)$; in that case, $R_{22}(r) = \overline{v'(x)v'(x+r)}$ and $R_{33}(r) = \overline{w'(x)w'(x+r)}$ are two *transversal correlations* that are generally not identical.

cross-correlations: this is the remaining correlation where the velocities in both points have different directions; for example, $R_{12}(r) = \overline{u'(x)v'(x+r)}$ with $\underline{r} = (r, 0, 0)$ is a *cross-correlation*.

The correlation functions provide insight in the spatial structure of turbulence. As an illustration we show in Fig. 9.4 some measurements of the longitudinal and transversal correlations in a turbulent channel flow. From these measurements it is clear that the length scale of the turbulence decreases as we approach the wall. This agrees with our discussion of the length scale in the inner layer in Chap. 6.

Because $R_{ij}(\underline{r})$ satisfies property *iii* in (9.16) we can again define a Fourier transform pair:

$$\phi_{ij}(\underline{\kappa}) = \frac{1}{8\pi^3} \int\!\!\!\int\!\!\!\int_{-\infty}^{+\infty} R_{ij}(\underline{r})e^{-i\underline{\kappa}\cdot\underline{r}}\, d\underline{r},$$

$$R_{ij}(\underline{r}) = \int\!\!\!\int\!\!\!\int_{-\infty}^{+\infty} \phi_{ij}(\underline{\kappa})e^{i\underline{\kappa}\cdot\underline{r}}\, d\underline{\kappa},$$

(9.17)

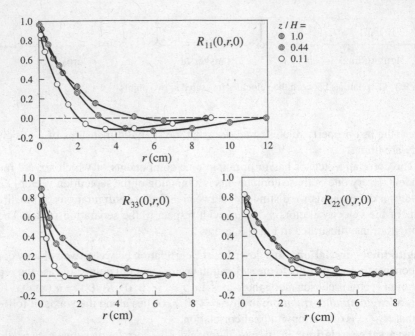

Fig. 9.4 Longitudinal and transversal spatial correlations in a channel flow. Measurements by G. Comte-Bellot using hot-wire anemometers. (Note that the distance from the wall is taken as the z-coordinate; see Fig. 2.2.) After: Townsend (1976)

where ϕ_{ij} is the *spectral tensor*. The theoretical background is that we can represent a homogeneous series as a sum of Fourier components. In this case these are waves with a *wave vector* $\underline{\kappa}$, which relates to the *wavelength* λ as

$$\underline{\kappa} = \frac{2\pi}{\lambda^2}\underline{\lambda}, \tag{9.18}$$

where: $\kappa = |\underline{\kappa}|$, and: $\underline{\kappa}/\kappa \equiv \underline{\lambda}/\lambda$ represents the *direction of propagation* of the wave. Using $\phi_{ij}(\underline{k})$ we can describe the *spatial structure* of homogeneous turbulence, where we can couple the wave number $|\underline{\kappa}|$ to the dimension of an eddy: a small $|\underline{\kappa}|$ relates to the *macrostructure*, while the *microstructure* is related to waves with large $|\underline{\kappa}|$. However, the complete spectral tensor $\phi_{ij}(\underline{\kappa})$ is in practice not easily handled. That is why we often look at simplified or reduced expressions.

First, we consider the so-called energy spectrum. The turbulent kinetic energy,

$$e = \frac{1}{2}\overline{u_i'^2} = \frac{1}{2}R_{ii}(0),$$

is related to the trace of the spectral tensor: $\frac{1}{2}\phi_{ii}(\underline{\kappa})$. However, we are only interested in the distribution of the energy in relation to the eddy size, that is $\kappa = |\underline{\kappa}|$, and not over the *direction* of $\underline{\kappa}$. We can eliminate this information about the direction by integrating $\frac{1}{2}\phi_{ii}(\underline{\kappa})$ over a spherical surface in $\underline{\kappa}$-space, with the result:

$$E(\kappa) = \frac{1}{2} \iint\limits_{|\underline{\kappa}|} \phi_{ii}(\underline{\kappa}) \, d\sigma, \qquad (9.19)$$

where the integration is carried out over a spherical surface $\kappa = |\underline{\kappa}|$ and where $d\sigma$ is a surface element of this spherical surface. On the basis of (9.17) and (9.19) it then follows that:

$$e = \int\limits_0^\infty E(\kappa) \, d\kappa. \qquad (9.20)$$

The term $E(\kappa)$ yields the distribution of energy over the various wave numbers κ, which is why $E(\kappa)$ is called the turbulent *energy spectrum*.

In Fig. 9.5, $E(\kappa)$ is illustrated schematically. We identify $\kappa_e \sim 1/\mathcal{L}$, where the spectrum has a maximum, with the large and energetic eddies. Close to $\kappa_d \sim 1/\eta$, the spectrum represents the microstructure. The region in between describes the energy *cascade process*. We return to the shape of the spectrum in this intermediate region in one of the following sections of this chapter.

The second simplification of the spectral tensor $\phi_{ij}(\underline{\kappa})$ is the so-called *one-dimensional spectrum*, where the separation vector $\underline{r}$ only varies along a given line. For example, if we take for this the x_1-axis, so that: $\underline{r} = (r_1, 0, 0)$, it follows for (9.17) that:

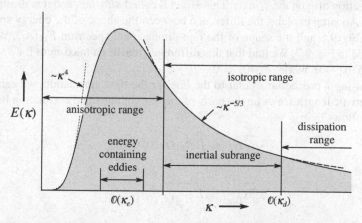

Fig. 9.5 The three-dimensional energy spectrum. The inertial subrange occurs roughly between $\kappa_e = 2\pi/\mathcal{L}$ and $\kappa_d = 2\pi/\eta$ and scales with $\kappa^{-5/3}$; see Sect. 9.5. The κ^4-scaling for $\kappa \to 0$ is described in Problem 6 of Sect. 9.6. The scaling for the dissipative range is discussed in Problem 12 of Sect. 9.6; see also Fig. 9.14

$$R_{ij}(r_1, 0, 0) = \int\limits_{-\infty}^{\infty} e^{i\kappa_1 r_1} \left\{ \iint\limits_{-\infty}^{\infty} \phi_{ij}(\kappa_1, \kappa_2, \kappa_3)\, d\kappa_2\, d\kappa_3 \right\} d\kappa_1.$$

This can be rewritten as:

$$R_{ij}(r_1, 0, 0) = \int\limits_{-\infty}^{\infty} e^{i\kappa_1 r_1} F_{ij}(\kappa_1)\, d\kappa_1, \tag{9.21}$$

where:

$$F_{ij}(\kappa_1) = \int\limits_{-\infty}^{\infty}\int\limits_{-\infty}^{\infty} \phi_{ij}(\kappa_1, \kappa_2, \kappa_3)\, d\kappa_2\, d\kappa_3, \tag{9.22}$$

is the *one-dimensional spectrum* and $R_{ij}(r_1)$ the *one-dimensional correlation function*. The terms $F_{ij}(\kappa_1)$ and $R_{ij}(r_1)$ are a *Fourier transform pair*, comparable with (9.10). Based on the orientation of the i and j-components with respect to the x_1-axis, we define $F_{11}(\kappa_1)$ as the *longitudinal spectrum* and $F_{22}(\kappa_1)$ and $F_{33}(\kappa_1)$ as the *transversal spectra*. The term $F_{ij}(\kappa_1)$ for $i \neq j$ is usually called the *co-spectrum*.

The term $F_{ij}(\kappa_1)$ describes the spatial structure along a line. The spatial structure is actually three-dimensional, so how should we interpret this? The answer lies in the definition (9.22) for $F_{ij}(\kappa_1)$, from which it follows that $F_{ij}(\kappa_1)$ consists of the integral over *all* contributions to the spectrum $\phi_{ij}(\underline{\kappa})$ with the x_1-component of the wavenumber $\underline{\kappa}$ equal to κ_1. In other words, in $F_{ij}(\kappa_1)$ we interpret *all* waves placed under an angle to the x_1-axis as waves *along* the x_1-axis with a wavenumber equal to the projection of $\underline{\kappa}$ on the x_1-axis. This effect is called *aliasing*, and it is illustrated in Fig. 9.6. Aliasing explains the difference between the shape of the energy spectrum $E(\kappa)$ in Fig. 9.5 and the shape of the one-dimensional spectrum $F_{11}(\kappa_1)$, which is illustrated in Fig. 9.7. We find that this difference reaches a maximum for a small κ, where: $E(0) \approx 0$, while: $F_{11}(0) \neq 0$.

Following a procedure similar to the one for the time correlation, we can define characteristic length scales on the basis of the one-dimensional correlation function. It thus follows that:

$$\overline{u'^2}\Lambda_L = \int\limits_{0}^{\infty} R_{11}(r_1, 0, 0)\, dr_1,$$

$$\overline{u'^2}\Lambda_T = \int\limits_{0}^{\infty} R_{22}(r_1, 0, 0)\, dr_1, \tag{9.23}$$

where we call Λ_L the *longitudinal* length scale and Λ_T *transversal* length scale. These are representative for the *macrostructure*. It should be clear that a second transversal length scale can be defined on the basis of R_{33}. We associate the microstructure to the second derivative of the correlation function at $r = 0$, that is:

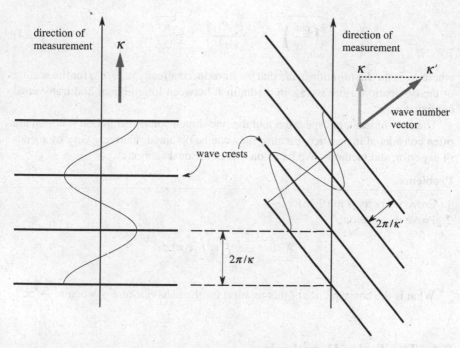

Fig. 9.6 Illustration of aliasing. No distinction can be made between a wave with wave number κ that travels in the direction of measurements and a wave with a wave number $\kappa' > \kappa$ that travels at an oblique angle with respect to the direction of measurement. After: Tennekes and Lumley (1972)

Fig. 9.7 The one-dimensional spatial spectrum $F_{11}(\kappa_1)$

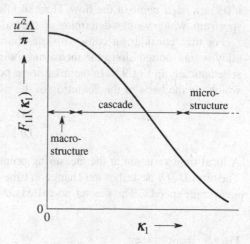

$$\overline{\left(\frac{\partial u'}{\partial x}\right)^2} \equiv -\left.\frac{\partial^2 R_{11}}{\partial r^2}\right|_{r=0} = \frac{\overline{u'^2}}{\lambda_T^2},$$

(9.24)

where λ_T is the *Taylor microscale* that we introduced already in (7.13) for the scaling of the dissipation. Here we again distinguish between longitudinal and transversal components.

The one-dimensional spectrum and the one-dimensional correlation function are often considered in practice, because they can be obtained relatively easy by means of experimental methods. We focus on this in the next section.

Problems

1. Prove property *ii* in (9.16).
2. Prove the relation:

$$\lambda^{-2} = \frac{2}{\overline{u'^2}} \int_0^\infty \kappa^2 F_{11}(\kappa)\, d\kappa.$$

What is the consequence of this relation for the behavior of F_{11} when $\kappa \to \infty$?

9.4 The Taylor Hypothesis

We assume an *Eulerian* measurement, which means that we measure a *time series* $u'(t)$ at a *fixed point* in the flow. How should we interpret the correlation and the spectrum, which we can determine on the basis of this measurement?

For this, consider an eddy with the characteristic macroscales $\mathcal{U}$ and $\mathcal{L}$. This eddy is transported along the measuring point with an average speed $\overline{u}$, as shown schematically in Fig. 9.8. In the measuring point we measure local time variations, which, on the basis of the definition of the material derivative (2.1), can be written as

$$\frac{\partial}{\partial t} = \frac{D}{Dt} - \overline{u}\frac{\partial}{\partial x}.$$

(9.25)

A local time variation at the measuring point is thus built up of two contributions. The term D/Dt describes the change in time while traveling along with the eddy at an average speed $\overline{u}$. The second, so-called *advection term*, describes the influence of

Fig. 9.8 Transport of an eddy along a fixed measuring point

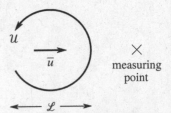

the spatial structure, which is transported along the fixed measuring point. We can estimate the order of magnitude of both terms as follows:

$$\frac{\partial}{\partial t} \approx \frac{\mathcal{U}}{\mathcal{L}} - \frac{\bar{u}}{\mathcal{L}} = \frac{\bar{u}}{\mathcal{L}}\left(\frac{\mathcal{U}}{\bar{u}} - 1\right),$$

where we take $\mathcal{L}/\mathcal{U}$ for the characteristic timescale of an eddy. The ratio $\mathcal{U}/\bar{u}$ is proportional to the so-called turbulence intensity and in most flows this is very small (typically less than 10%). From this it follows that the first term in (9.25) is negligible compared to the second term, or:

$$\frac{\partial}{\partial t} \approx -\bar{u}\frac{\partial}{\partial x}. \tag{9.26}$$

We can interpret this relation as follows. The time it takes for an eddy to advect past the measuring point by the average velocity is much shorter than the time that it takes for the eddy to change its shape. This is referred to as the hypothesis of '*frozen turbulence*', or *Taylor hypothesis* (Taylor 1938).

The measured fluctuations are thus only determined by the advection of the spatial structure of the flow along the measuring point. This means that we should not couple the correlation function and the spectrum of the measured time series to the time structure of the turbulence, but to the *spatial* structure. In other words, the measured time spectrum $S(\omega)$ should be interpreted as the one-dimensional spatial spectrum $F(\kappa)$, of which we saw in the previous section that it describes the spatial structure of the turbulence along a line. Here, the relation between ω and κ is given by:

$$\kappa = \frac{\omega}{\bar{u}}.$$

It thus follows that:

$$F(\kappa)\,d\kappa = F\left(\frac{\omega}{\bar{u}}\right)\frac{d\omega}{\bar{u}} = S(\omega)\,d\omega \quad \Rightarrow \quad S(\omega) = \frac{1}{\bar{u}}F\left(\frac{\omega}{\bar{u}}\right).$$

In short, Eulerian measurements in a single point give *no* information whatsoever about the *temporal* structure of turbulence, but instead about the *spatial* structure along a line that is parallel to the average transport velocity $\bar{u}$. In order to determine the true temporal structure of turbulence, we have to resort to the *Lagrangian* measurements.

Problems

1. Define the Lagrangian (L) and Eulerian (E) Taylor micro-timescales as:

$$\overline{\left(\frac{Du'}{Dt}\right)^2} \sim \left(\frac{\mathcal{U}}{\lambda_T^{(L)}}\right)^2,$$

$$\overline{\left(\frac{\partial u'}{\partial t}\right)^2} \sim \left(\frac{\mathcal{U}}{\lambda_T^{(E)}}\right)^2,$$

respectively.

(a) Show that, at the same time, the following holds:

$$\overline{\left(\frac{Du'}{Dt}\right)^2} \sim \left(\frac{\upsilon}{\tau}\right)^2,$$

where υ and τ are the *Kolmogorov scales*. Such a scaling does not hold for $\overline{(\partial u'/\partial t)^2}$. Why not?

(b) Derive a relation between the integral timescale $\mathcal{T}$ and $\lambda_T^{(L)}$, and for $\lambda_T^{(L)}$ and the Kolmogorov timescale τ.

(c) Show that $\lambda_T^{(E)}/\lambda_T^{(L)}$ is a function of a Reynolds number, from which it follows that $\lambda_T^{(E)} \ll \lambda_T^{(L)}$. Give an interpretation of this result.

(d) Consider a turbulent flow with $\bar{u} = 0$. Do the results above have to be adjusted?

2. We consider a Lagrangian correlation and the associated Lagrangian spectrum $S(\omega)$. Prove that

$$\frac{\overline{u'^2}}{\lambda_T^{(L)2}} = \int_{-\infty}^{\infty} \omega^2 S(\omega)\, d\omega.$$

Next, consider the spectrum

$$S(\omega) = \frac{\overline{u'^2}\mathcal{T}}{\pi} \frac{e^{-\tau^2 \omega^2}}{1 + \omega^2 \mathcal{T}^2},$$

where τ is a timescale.

Using this spectrum, calculate the energy and $\lambda_T^{(L)}$. Make use of the following integrals:

$$\int_0^{\infty} \frac{e^{-\mu^2 x^2}}{x^2 + \beta^2}\, dx = [1 - \Phi(\beta\mu)] \frac{\pi}{2\beta} e^{\beta^2 \mu^2},$$

$$\int_0^{\infty} \frac{x^2 e^{-\mu^2 x^2}}{x^2 + \beta^2}\, dx = \frac{\sqrt{\pi}}{2\mu} - \frac{\pi}{2} \beta e^{\beta^2 \mu^2} [1 - \Phi(\beta\mu)],$$

where $\Phi(s)$ represents the *error function*:

$$\Phi(s) = \frac{2}{\sqrt{\pi}} \int\limits_0^s e^{-t^2}\, dt.$$

Show that, under the condition that: $\lambda_T^{(L)} \ll \mathcal{T}$, τ is proportional to the Kolmogorov timescale.

What does in this case the correlation function look like?

9.5 Scaling of Turbulence Spectra

We argued that we can interpret the spectra $\phi_{ij}(\underline{\kappa})$, $E(\kappa)$ and $F_{11}(\kappa_1)$ in terms of the *spatial structure* of turbulence. We already discussed this spatial structure extensively in another context; in particular, we found that there is a *macrostructure* and a *microstructure*, which are dynamically decoupled, and therefore each have their own scaling parameters. Now what is the consequence of this for the turbulence spectrum?

We focus on the one-dimensional spectrum $F_{11}(\kappa)$, although the arguments are in principle valid for the other spatial spectra as well. Based on the scaling of the macrostructure mentioned above, it is obvious to describe the spectrum in this region as:

$$F_{11}(\kappa) = \mathcal{L}\mathcal{U}^2\, \phi_e(\kappa\mathcal{L}), \tag{9.27}$$

which is thus valid for the region around $\kappa \approx \kappa_e = 2\pi/\mathcal{L}$. Here, $\mathcal{U}$ and $\mathcal{L}$ are the characteristic scales of the *macrostructure*. The term ϕ_e describes the shape of the spectrum in the macroscale range for the larger eddies. This function is basically different for every flow geometry, because we know that the large eddies are determined by this geometry.

For the spectrum in the region of the *microstructure*, that is close to $\kappa \approx \kappa_d = 2\pi/\eta$, it follows that:

$$F_{11}(\kappa) = \eta\, \upsilon^2\, \phi_d(\kappa\eta), \tag{9.28}$$

where η and υ are the Kolmogorov length scale and velocity scale, respectively. The term ϕ_d describes the shape of the spectrum in the microscale range. Based on this scaling it follows that (9.28) does not depend on the macroscales, and thus not on the flow geometry. This leads to the important conclusion that ϕ_d is *universal*, or the structure of the microscales is the same for *all* turbulent flows. This is known as the *universal equilibrium theory* of Kolmogorov (1991, 1962). For additional support of this theory we refer to Sect. 8.3. There we showed, based on a first-order approximation of the enstrophy equation, that the dynamics of the microstructure are determined by the local equilibrium between *production* and *destruction* of enstrophy, *independent* of the macrostructure.

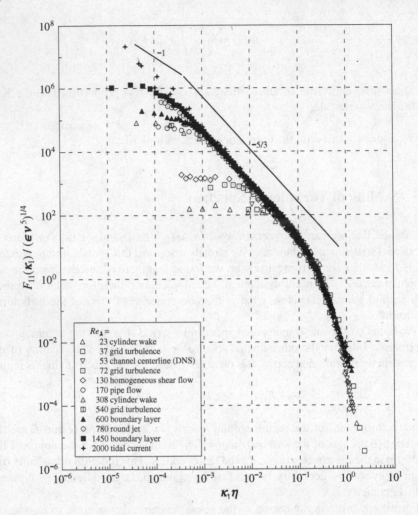

Fig. 9.9 One-dimensional spectra scaled with respect to the microstructure from various turbulent flows. This demonstrates the universal character of the microstructure and illustrates the '−5/3' behavior of the inertial subrange. Data of: cylinder wake (Uberoi and Freymuth 1969); grid turbulence (Kistler and Vrebalovich 1966; Comte-Bellot and Corrsin 1971); DNS of channel flow (Kim and Antonia 1993); homogeneous shear flow (Champagne et al. 1970); pipe flow (Laufer 1954); boundary layer (Saddoughi and Veeravalli 1994); round jet (Gibson 1963); tidal current (Grant et al. 1962). Re_λ is the Reynolds number based on the Taylor microscale. After: Saddoughi and Veeravalli (1994)

In Fig. 9.9 we show experimental and numerical data that provide a validation of this universal character of the microstructure. Several turbulence spectra are shown, scaled according to (9.28) with $\eta\upsilon^2 = (\epsilon\nu^5)^{1/4}$, plotted against $\kappa_1\eta$. Despite the fact that the flows occur in very different geometries and under different conditions, the spectra in the region of large wave numbers are nearly identical.

We have also seen in Sect. 8.3 that this local equilibrium for the microstructure is only valid for large values of the Reynolds number $\mathcal{U}\mathcal{L}/\nu$. In terms of the spectrum this means that the regions in the spectrum that can be associated with the macrostructure and the microstructure are well separated, or: $\kappa_d/\kappa_e \gg 1$. Now let us turn our attention to the intermediate region, where the *energy cascade process* occurs. This region is called the *inertial subrange*, and the spectrum in this range does not depend on the macroscales, because for this intermediate region it holds that $\kappa \gg \kappa_e$. On the other hand, the spectrum cannot depend on ν either, because: $\kappa \ll \kappa_d$. Based on this the following shape for the inertial subrange is found using dimensional analysis:

$$F_{11}(\kappa) = \alpha\epsilon^{2/3}\kappa^{-5/3}, \qquad (9.29)$$

where α is a universal constant, which is called the *Kolmogorov constant*. From experimental data it is found that: $\alpha \approx 0.26$.

The result (9.29) can also be found based on a formal *matching* between (9.27) and (9.28) in a manner similar to the approach that we applied in Sect. 6.3 to derive the logarithmic velocity profile. The matching condition for the inertial subrange can be formulated as follows:

$$\lim_{\kappa\mathcal{L}\to\infty} \mathcal{L}\mathcal{U}^2\phi_e(\kappa\mathcal{L}) = \lim_{\kappa\eta\to 0} \eta\,\upsilon^2\phi_d(\kappa\eta).$$

If we apply the relation $\epsilon = \mathcal{U}^3/\mathcal{L} = \upsilon^3/\eta$ to substitute for $\mathcal{U}$ and υ, respectively, and then divide both sides by $\epsilon^{2/3}\kappa^{-5/3}$, it follows that

$$\lim_{\kappa\mathcal{L}\to\infty} \frac{\phi_e(\kappa\mathcal{L})}{(\kappa\mathcal{L})^{-5/3}} = \lim_{\kappa\eta\to 0} \frac{\phi_d(\kappa\eta)}{(\kappa\eta)^{-5/3}}.$$

This equality can only be satisfied when both limits equal the same constant α.

Equation (9.29) represents one of the most important foundations of current turbulence theory, and many measurements were performed to verify this result. In Fig. 9.9 we already saw several experimental and numerical (DNS) data that indeed verify the existence of this $-5/3$-behavior. Furthermore, it can be noted that for increasing Reynolds numbers the range increases over which (9.29) is valid. This relates of course to the increasing separation between the macrostructure and the microstructure with increasing Reynolds number.

So far, we limited ourselves to one-dimensional spectra. The inertial subrange can also be derived for the turbulent energy spectrum defined in (9.19). It follows that:

$$E(\kappa) = \beta\epsilon^{2/3}\kappa^{-5/3}, \qquad (9.30)$$

where the constant β is approximately 1.6. In case of *isotropic turbulence* a relation exists between α and β, and for this we refer to Problem 1 of Sect. 9.6.

The $-5/3$-behavior in the inertial subrange is a very important result, because it implies an indirect validation of our picture of turbulence in terms of a macrostructure and a microstructure. At the same time, we can consider this as a confirmation of all our previous findings that are all based on this presumed picture of the structure of turbulence. At the same time, the universality of the microstructure provides a strong rationale for the method of *large-eddy simulation* (LES) described in Sect. 8.6.

Problems

1. Explain why the range where (9.29) is valid varies for the different flows depicted in Fig. 9.9.
2. Show that for an Eulerian time spectrum in the inertial subrange we have: $S(\omega) = \alpha(\overline{u}\,\epsilon)^{2/3}\omega^{-5/3}$, while for the Lagrangian time spectrum in this region we have: $S(\omega) = \beta_L \epsilon \omega^{-2}$. (For isotropic turbulence: $\beta_L \approx 0.8$).
3. Define $D_{11}(r) = \{u'(x) - u'(x+r)\}^2$ as the *longitudinal structure function*. What is the relation between D_{11} and the correlation function R_{11}? Discuss the physical meaning of D_{11}, and determine on the basis of this an expression for D_{11} in the inertial subrange. (See also Problem 2 of Sect. 10.1.)
4. Consider the following model spectrum:

$$F_{11}(\kappa) = \begin{cases} F(0) & \text{for: } \kappa < \kappa_e \\ \alpha\epsilon^{2/3}\kappa^{-5/3} & \text{for: } \kappa_e \leqslant \kappa \leqslant \kappa_d \\ 0 & \text{for: } \kappa > \kappa_d \end{cases}$$

   Based on this spectrum, determine:

   (a) The relation between κ_e and Λ_L.
   (b) The relation between ϵ, $\overline{u'^2}$, and Λ_L.

5. An energy spectrum with the following shape is given:

$$E(\kappa) = \begin{cases} A\kappa^m & \text{for: } \kappa < \kappa_e, \\ \beta\epsilon^{2/3}\kappa^{-5/3} & \text{for: } \kappa > \kappa_e, \end{cases}$$

   where β is the Kolmogorov constant for the energy spectrum. Using this spectrum, calculate the decay of homogeneous turbulence. First determine a relation between κ_e and ϵ given that the spectrum must be continuous at $\kappa = \kappa_e$. Next, determine a relation between the total kinetic energy e and the dissipation ϵ. Substitute this relation in the kinetic energy equation for homogeneous turbulence. Based on the solution, show how e and κ_e vary as a function of time.

6. Consider a wall-bounded turbulent flow, such as a boundary layer flow, where the size of the large-scale eddies is limited by the distance from the wall, as described in Sect. 6.2. Use the *matching* approach described in this section to show that

$$F_{11}(\kappa) \sim \kappa^{-1} \quad \text{for: } \frac{1}{\delta} < \kappa < \frac{1}{\ell},$$

with: $\ell \approx ky$, and where δ is the boundary-layer thickness. This -1 scaling behavior can also be observed in Fig. 9.9; further experimental data are given by Perry et al. (1986).

7. Suppose that the dissipation ϵ_0 is not distributed uniformly in space, but that it is *intermittent* according to a *fractal structure*, as illustrated in Fig. 4.8 (see also Fig. 8.11). This implies that, when we divide the volume $V_0 = \mathcal{O}(\ell_0^3)$ (with: $\ell_0 \sim \mathcal{L}$) into sub-volumes $V_n = \mathcal{O}(\ell_n^3)$, the dissipation is *not* identical zero only in a fraction χ of all sub-volumes V_n in V_0. The representative value for the dissipation in this volume $\Delta V_n = \chi V_0$ is ϵ_n.

We can now write:

$$\epsilon_0 \equiv \langle \epsilon \rangle = \left(\frac{\Delta V_n}{V_0} \right) \epsilon_n = \left(\frac{\ell_n}{\ell_0} \right)^{3-D} \epsilon_n,$$

where D is a *fractal dimension*. (When $D = 3$, the dissipation is distributed uniformly in space.) Here, $\langle \cdots \rangle$ indicates a spatial averaging. (Note that $\langle \epsilon^a \rangle = \langle \epsilon \rangle^a$ only holds when the dissipation is distributed uniformly in space.)

For ϵ_n we can write: $\epsilon_n \sim v_n^3/\ell_n$, where v_n is a characteristic velocity in a volume with a dimension ℓ_n.

(a) The dissipation eventually occurs in the microstructure ℓ_η. Suppose that this microstructure is characterized by $\epsilon_\eta = v_\eta^3/\ell_\eta = \nu v_\eta^2/\ell_\eta^2$. Derive a relation for ℓ_η/ℓ_0 as a function of the Reynolds number. Consider the Kolmogorov scale, defined as: $\eta = \nu^{3/4}/\langle \epsilon^{1/4} \rangle$. Derive a relation between ℓ_η and η as a function of the Reynolds number.

(b) We can write the contribution to the energy at the scale ℓ_n as:

$$e_n = \left(\frac{\ell_n}{\ell_0} \right)^{3-D} v_n^2.$$

Define the spectrum when $E(\kappa_n) \sim e_n \ell_n$, with $\kappa_n \sim \ell_n^{-1}$. Use this to determine the spectrum as a function of κ and compare the result with (9.29).

(c) Argue that, on the basis of Kolmogorov scaling, it follows for the microstructure that:

$$\overline{\left(\frac{\partial u'}{\partial x} \right)^n} \sim \nu^{-n/2} \overline{\epsilon^{n/2}}.$$

Using the model above, calculate the *skewness*

$$S = \frac{\overline{(\partial u'/\partial x)^3}}{\left\{ \overline{(\partial u'/\partial x)^2} \right\}^{3/2}},$$

and *kurtosis*

$$K = \frac{\overline{(\partial u'/\partial x)^4}}{\left\{\overline{(\partial u'/\partial x)^2}\right\}^2},$$

of the velocity gradients as a function of the Reynolds number $Re = \mathcal{U}\mathcal{L}/\nu$.

8. Consider *homogeneous turbulence* with temperature fluctuations. The temperature field is also homogeneous.

Define the correlation function:

$$R_\theta(\underline{r}) = \overline{\theta'(\underline{x})\theta'(\underline{x}+\underline{r})}.$$

Then, the spectrum $F_\theta(\underline{\kappa})$ of the temperature fluctuations is defined as:

$$R_\theta(\underline{r}) = \iiint F_\theta(\underline{\kappa}) e^{i\underline{\kappa}\cdot\underline{r}} \, d\underline{\kappa},$$

$$F_\theta(\underline{\kappa}) = \frac{1}{8\pi^3} \iiint R_\theta(\underline{r}) e^{-i\underline{\kappa}\cdot\underline{r}} \, d\underline{r}.$$

Define $E_\theta(\kappa)$, with $\kappa = |\underline{\kappa}|$, as the integral of $\frac{1}{2}F_\theta(\underline{\kappa})$ over a spherical surface in κ-space with a radius κ.

(a) Show that:

$$\frac{1}{2}\overline{\theta'^2} = \int\limits_0^\infty E_\theta(\kappa) \, d\kappa,$$

and give an interpretation of E_θ on the basis of this result.

(b) Consider the *Prandtl number* $Pr = \nu/\alpha = 1$, where α is here the *thermal diffusivity*. Deduce an explicit expression for $E_\theta(\kappa)$ in the inertial subrange.

(c) Consider $Pr = \nu/\alpha \ll 1$. Argue that the *inertial subrange* for the *temperature spectrum* is smaller than the inertial subrange for the velocity energy spectrum. Do this by making use of the *Corrsin scales* η_θ, which are derived in Problem 2 of Sect. 10.2.

(d) Consider $Pr = \nu/\alpha \gg 1$. Argue that the temperature spectrum is *wider* than the velocity spectrum, and show that for $\kappa > 1/\eta$ (where η is the Kolmogorov scale) it holds that:

$$E_\theta(\kappa) \sim \kappa^{-1}.$$

This is called the *Batchelor spectrum*, and the range where this spectrum is valid is called the *viscous convective subrange*.

(e) Make sketches of the spectrum $E_\theta(\kappa)$ for various values of the Prandtl number.

9.6 Isotropic Turbulence

To conclude this chapter we take a closer look at a specific type of turbulence, namely *homogeneous isotropic turbulence*. In addition to homogeneity, which means that the statistical properties of this type of turbulence are invariant with respect to *translation*, isotropy implies that the statistical properties are also invariant to *rotation* and *reflections* of the coordinate system. (It should be noted that isotropy of turbulence implies that the turbulence must also be homogeneous; we therefore simply refer to *isotropic turbulence* is the remainder of this section.)

Before we discuss the theory of isotropic turbulence, we must first question how realistic this type of turbulence is. In Chap. 7 we saw that the production of turbulent energy implies that the macrostructure of turbulence is anisotropic. Thus, the model of isotropic turbulence is definitely *not* applicable to the macrostructure. The situation is different for the microstructure. During the cascade process, with every step, information on the anisotropy of the production processes is lost. The *microstructure* that lies at the end of the cascade process thus has lost any preferred direction, which means that the microstructure is *isotropic*. Recall that because of its isotropic character, the microstructure has universal properties; it is therefore relevant to consider the special case of isotropic turbulence. Please note that this is of particular relevance to defining suitable sub-grid closure models for large-eddy simulation (see Sect. 8.6).

Above, we defined isotropic turbulence as invariant to rotations and reflections. This immediately implies that the variance of the velocity fluctuations is identical in all directions:

$$\overline{u'^2_\alpha} = \frac{2}{3}e, \tag{9.31}$$

with *no summation* over the index α. Moreover, we can simplify the correlation tensor R_{ij}. Based on invariance with respect to reflections of the coordinate system it follows that:

$$R_{ij}(\underline{r}) = R_{ji}(-\underline{r}) = R_{ji}(\underline{r}),$$

or, in other words, that the correlation tensor is *symmetric*. Next, we show that *all* cross correlations (as defined in Fig. 9.3) are equal to zero. For example, consider $R_{13}(\underline{r})$, with: $\underline{r} = (r, 0, 0)$, as illustrated in Fig. 9.10; based on a rotation around the x_1-axis in this figure it follows that:

$$R_{13}(r) = -R_{13}(r) = 0.$$

Fig. 9.10 The cross correlation $R_{13}(r)$ for isotropic turbulence; due to rotational symmetry: $R_{13}(r) = -R_{13}(r) = 0$

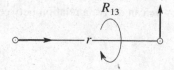

The only correlations that do not vanish are the longitudinal correlation and the transversal correlation, which are indicated by the functions $f(r)$ and $g(r)$, respectively. As a consequence of the rotational symmetry, $f(r)$ and $g(r)$ are *independent* of the direction, and thus only a function of $r = |\underline{r}|$. Also, it is easy to see that there is only a *single* transversal correlation function. Based on these functions we can introduce a *longitudinal integral length scale* Λ_L and *transversal integral length scale* Λ_T:

$$\overline{u'^2}\Lambda_L = \int_0^\infty f(r)\,dr, \quad \text{and:} \quad \overline{u'^2}\Lambda_T = \int_0^\infty g(r)\,dr. \tag{9.32}$$

The functions $f(r)$ and $g(r)$ can be utilized to define also longitudinal and transversal Taylor micro scales:

$$\lambda_L^2 = \overline{u'^2} \left/ \overline{\left(\frac{\partial u'}{\partial x}\right)^2} \right. = -\overline{u'^2} \left/ \left.\frac{\partial^2 f}{\partial r^2}\right|_{r=0} \right. ,$$

$$\lambda_T^2 = \overline{v'^2} \left/ \overline{\left(\frac{\partial v'}{\partial x}\right)^2} \right. = -\overline{v'^2} \left/ \left.\frac{\partial^2 g}{\partial r^2}\right|_{r=0} \right. . \tag{9.33}$$

We can now completely describe the correlation tensor between two random velocity components using the functions $f(r)$ and $g(r)$. The result reads:

$$R_{ij}(\underline{r}) = \{f(r) - g(r)\}\frac{r_i r_j}{r^2} + g(r)\delta_{ij}. \tag{9.34}$$

This equation represents a decomposition of the general correlation R_{ij} into a longitudinal component and a transversal component, that is in projections that are perpendicular and parallel to the direction of $\underline{r}$. We can completely describe the structure of the isotropic velocity field with (9.34). Instead of the nine components in the general correlation tensor R_{ij} we now only need *two* functions.

So far we have limited ourselves to a geometric description of the *structure* of isotropic turbulence. Now we require that (9.34) satisfies the equations of motion. First, we take the continuity equation, which states that $\partial u_i/\partial x_i = 0$. When we apply this condition to the correlation function R_{ij} it follows that: $\partial R_{ij}/\partial r_i = 0$. Using the relations:

$$\frac{\partial}{\partial r_i} = \frac{r_i}{r}\frac{\partial}{\partial r}, \quad \frac{\partial r_j}{\partial r_i} = \delta_{ij}, \quad \text{and:} \quad \frac{\partial r_i}{\partial r_i} = 3,$$

we can derive a relation between $f(r)$ and $g(r)$, which reads:

$$r\frac{\partial f}{\partial r} + 2(f - g) = 0. \tag{9.35}$$

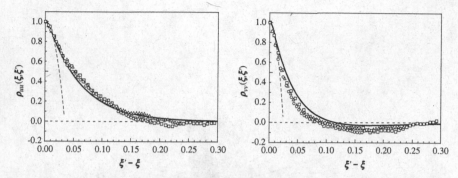

Fig. 9.11 The longitudinal (*left*) and transversal (*right*) correlations in the central region of a turbulent jet at three distances from the jet nozzle: $x/d = 53$ ($\square$), 72 ($\triangle$), and 80 (o). The spatial correlations $R_{ij}(x' - x)$ in the jet become self-similar when plotted as a function of $\xi' - \xi$, with: $\xi = \ln(x/d)$, and: $\xi' = \ln(x'/d)$ (Ewing et al. 2007). The *solid lines* represent fitted functions for f and g, which satisfy (9.35). The *dashed lines* correspond to the Taylor scaling in (9.24), with: $\lambda_T^{(L)}/\lambda_T^{(T)} = \sqrt{2}$. Experimental data from: Fukushima et al. (2002)

This relation means that we can describe the structure of isotropic turbulence with a *single* scalar function $f(r)$. Figure 9.11 shows the longitudinal and transversal spatial correlations in the central region of a turbulent jet. The correlation $f(r)$ has been fitted to the experimental data, and then (9.35) is used to determine $g(r)$, which is compared in the right graph with the experimental data for the transversal correlation.

Figure 9.12 shows a reconstruction of the conditionally averaged flow field for a given velocity fluctuation (represented by the large arrow); the conditional velocity field is computed by means of *linear stochastic estimation* (LSE), which estimates the most likely flow field on the basis of $R_{ij}(r)$ (Adrian, 1979). The reconstruction in Fig. 9.12 is based on (9.34) and (9.35) for given $f(r)$.

Next, we define the *Fourier transforms* of $f(r)$ and $g(r)$ as the *longitudinal spectrum* $F(\kappa)$ and the *transversal spectrum* $G(\kappa)$:

$$F(\kappa) = \frac{1}{2\pi} \int_{-\infty}^{\infty} f(r)e^{-i\kappa r}\, dr, \quad \text{and:} \quad G(\kappa) = \frac{1}{2\pi} \int_{-\infty}^{\infty} g(r)e^{-i\kappa r}\, dr. \tag{9.36}$$

In addition, we introduce the *energy spectrum* $E(\kappa)$, which is defined according to (9.19) and (9.20). For isotropic turbulence we can calculate the relation between these spectra explicitly; the derivation is found in Problem 1, and the result reads:

$$E(\kappa) = \kappa^3 \frac{d}{d\kappa}\left(\frac{1}{\kappa}\frac{dF}{d\kappa}\right), \quad \text{with:} \quad G(\kappa) = \frac{1}{2}F(\kappa) - \frac{1}{2}\kappa\frac{dF}{d\kappa}. \tag{9.37}$$

With these equations we can convert the various spectra into each other.

We argued above that isotropic turbulence applies to the microstructure. We also know that the viscous dissipation ϵ takes place at this microstructure. It is therefore

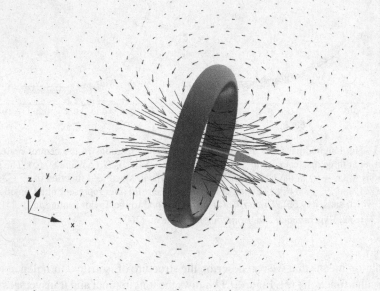

Fig. 9.12 Conditional velocity field for a velocity fluctuation (*large arrow* in the center parallel to the x-axis) in isotropic turbulence computed using a *linear stochastic estimate* (Adrian 1979) based on $R_{ij}(r)$ given (9.34), with arrows shown in the (x, y)-plane. The structure is rotationally symmetric with respect to the x-axis. The conditional flow pattern has the shape of a *vortex ring* with a radius of about $2\Lambda_L$

obvious to use the results for isotropic turbulence to calculate ϵ. By definition it holds that:

$$\epsilon = \nu \overline{\left(\frac{\partial u_i'}{\partial x_j}\right)^2}.$$

Using the definition in (9.17) of the spectral tensor ϕ_{ij} and the definitions (9.19) and (9.20) for $E(\kappa)$, it follows for ϵ that:

$$\epsilon = -\nu \left.\frac{\partial^2 R_{ii}}{\partial r_j^2}\right|_{r=0} = 2\nu \int_0^\infty \kappa^2 E(\kappa)\, d\kappa. \tag{9.38}$$

When we apply (9.37) to this expression, it follows that:

$$\epsilon = 30\,\nu \int_0^\infty \kappa^2 F(\kappa)\, d\kappa = 15\,\nu \overline{\left(\frac{\partial u'}{\partial x}\right)^2}. \tag{9.39}$$

This means that we can calculate ϵ using the derivative of only *one* velocity component. This equation is very useful in case we perform measurements in a single point (that is, Eulerian measurements). Using the Taylor hypothesis, we can now calculate $\partial u'/\partial x$ as:

$$\overline{\left(\frac{\partial u'}{\partial x}\right)^2} = \overline{\left(\frac{\partial u'}{\partial t}\right)^2} \Big/ \overline{u}^2,$$

where $\partial u'/\partial t$ follows from differentiation of the measured time series.

So far we only considered the kinematics of isotropic turbulence. We conclude this section by taking also into consideration the Navier–Stokes equations, that is the *dynamic* properties of the turbulence. On the basis of the Navier–Stokes equations we can derive an equation that describes $f(r)$ as a function of time. The result reads:

$$\frac{\partial}{\partial t}\left(\overline{u'^2}\, r^4 f\right) - \frac{\partial}{\partial r}\left\{r^4 k(r)\right\} = 2\nu\overline{u'^2}\frac{\partial}{\partial r}\left(r^4\frac{\partial f}{\partial r}\right). \tag{9.40}$$

This expression is known as the *von Kármán–Howarth equation*. For details on its derivation we refer to the book by Hinze (1975). We see that a new unknown function $k(r)$ emerges in this equation; this is a *third-order longitudinal correlation*, defined as:

$$k(r) = \overline{u'(x)u'(x)u'(x+r)}, \tag{9.41}$$

which is illustrated in Fig. 9.13, and it has the following properties:

$$k(0) = \overline{u'^3} = 0, \quad k(r) = -k(-r), \quad \text{and:} \quad k(r) = \frac{1}{6}\overline{\left(\frac{\partial u'}{\partial x}\right)^3}r^3 \quad \text{for: } r \to 0,$$
$$\tag{9.42}$$

which can be proven using the rotational and reflectional symmetries of isotropic turbulence (see also Problem 2 below).

The most important conclusion following from (9.40) is that again we encounter a closure problem; we find a new unknown function $k(r)$ in the equation for $f(r)$, and solving (9.40) is only possible when we specify a closure relation for $k(r)$. However, to this date a suitable closure relation that would be valid under general conditions has not been found.

Problems

1. Define E_1 and E_2 as:

Fig. 9.13 The third-order correlation $k(r)$

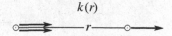

$$E_1 = \frac{2}{\pi} \int\limits_0^\infty f(r) \cos(\kappa r)\, dr, \quad \text{and:} \quad E_2 = \frac{2}{\pi} \int\limits_0^\infty g(r) \cos(\kappa r)\, dr.$$

(a) Show that: $E_1(k) = 2F(k)$.

(b) In the inertial subrange it follows that:

$$E(\kappa) = \beta \epsilon^{2/3} \kappa^{-5/3},$$
$$F(\kappa) = \alpha \epsilon^{2/3} \kappa^{-5/3},$$
$$E_1(\kappa) = \alpha' \epsilon^{2/3} \kappa^{-5/3},$$

where, based on experimental data, it follows that: $\beta \approx 1.5\text{--}1.7$. Prove that: $\alpha = \frac{9}{55}\beta$, and that: $\alpha' = \frac{18}{55}\beta$.

(c) Prove that in the inertial subrange it holds that:

$$G(\kappa) = \frac{4}{3} F(\kappa).$$

2. Prove that:

$$k(r) = \frac{1}{6} \overline{\left(\frac{\partial u'}{\partial x}\right)^3} r^3,$$

for: $r \to 0$. (Hint: consider $K(r) = \overline{\{u'(x) - u'(x+r)\}^3}$, and expand this in a Taylor series.)

3. Prove, on the basis of (9.35), that:

$$\Lambda_T = \frac{1}{2} \Lambda_L,$$

and that:

$$\int\limits_0^\infty r\, g(r)\, dr = 0.$$

Show, on the basis of this last relation, that $g(r)$ has to be *negative* somewhere. Discuss the *shape* of $f(r)$ and $g(r)$, as illustrated in Fig. 9.11, based on this last relation.

4. Prove that: $\lambda_T = \lambda_L/\sqrt{2}$. Using this, derive Eq. (9.39).

5. Based on symmetry and invariance properties, it follows for the spectrum of isotropic turbulence that:

$$\phi_{ij} = \frac{\kappa_i \kappa_j}{\kappa^2} \phi^{(L)}(\kappa) + \left(\delta_{ij} - \frac{\kappa_i \kappa_j}{\kappa^2}\right) \phi^{(T)}(\kappa),$$

with: $\kappa = (\kappa_i \kappa_i)^{1/2}$, and where $\phi^{(L)}$ and $\phi^{(T)}$ are called the *longitudinal* and *transversal* components.

(a) Based on the continuity equation, derive that:

$$\phi^{(L)}(\kappa) = 0,$$

which means that the spectrum ϕ_{ij} is orthogonal to the vector $\underline{\kappa}$.

(b) Next, using Eq. (9.19), derive that:

$$\phi_{ij} = \left(\delta_{ij} - \frac{\kappa_i \kappa_j}{\kappa^2} \right) \frac{E(\kappa)}{4\pi\kappa^2}.$$

(c) Now consider the special case in which $\underline{r} = (r_1, 0, 0)$ and $\underline{\kappa} = (\kappa_1, 0, 0)$. Using the definition in (9.21) for $R_{11}(r_1)$, it follows that:

$$F(\kappa_1) = \int_{\kappa_1}^{\infty} \left(1 - \frac{\kappa_1^2}{\kappa^2} \right) \frac{E(\kappa)}{2\kappa} \, d\kappa.$$

Use this expression to derive the relation between $E(\kappa)$ and $F(\kappa)$ in (9.37).

6. Based on the relation in (9.37) and the properties of $F(\kappa)$ for $\kappa \to 0$, derive that:

$$E(\kappa) = \frac{1}{3\pi} B \kappa^4,$$

for: $\kappa \to 0$. Here, B is given by:

$$B = \int_{0}^{\infty} r^4 f(r) \, dr,$$

and is known as *Loitsyanskiy's invariant*.

Derive, on the basis of the Von Kármán–Howarth equation in (9.40), the condition under which the Loitsyanskiy's invariant is *constant*, that is: $B \neq f(t)$.

7. Based on a series expansion of the Von Kármán–Howarth equation, derive the following expressions for $r \to 0$:

$$\frac{\partial e}{\partial t} = -\epsilon,$$

$$\frac{\partial \overline{\omega'^2}}{\partial t} = -35 \overline{\left(\frac{\partial u'}{\partial x} \right)^3} - 70\nu \overline{\left(\frac{\partial^2 u'}{\partial x^2} \right)^2}.$$

Interpret these expressions and describe the physical processes represented by the various terms.

8. Consider decaying isotropic turbulence. Assume that the longitudinal correlation function $f(r)$ can be described by the following similarity equation:

$$f(r) = \mathcal{U}^2 F\left(\frac{r}{\mathcal{L}}\right),$$

where the scales $\mathcal{U}$ and $\mathcal{L}$ are functions of t.

(a) Derive, on the basis of the kinetic energy equation and Loitsyanskiy's invariant how $\mathcal{U}$ and $\mathcal{L}$ vary as a function of t.

(b) Apply the similarity hypothesis to the complete Von Kármán–Howarth equation in (9.40), so that $k(r) = \mathcal{U}^3 K(r/\mathcal{L})$. How do $\mathcal{U}$ and $\mathcal{L}$ vary as a function of t in this case?

9. We argued that isotropic turbulence is a valid model for the microstructure only. It is therefore not convenient to work with the correlation function. Explain why.

Instead, we can use the *longitudinal structure function*, defined as: $D(r) = \overline{\{u'(x+r) - u'(x)\}^2}$. Derive on the basis of the Von Kármán–Howarth equation the following expression for $D(r)$:

$$\frac{\partial D}{\partial t} + \frac{1}{3}\frac{1}{r^4}\frac{\partial}{\partial r}\left(r^4 D^{3/2} S\right) - 2\nu\frac{1}{r^4}\frac{\partial}{\partial r}\left(r^4\frac{\partial D}{\partial r}\right) = -\frac{4}{3}\epsilon, \quad \text{with:} \quad S = \frac{\overline{\{u'(x+r) - u'(x)\}^3}}{D^{3/2}}.$$

Argue that, in the *inertial subrange*, the first term and the last term at the left-hand side of this equation are negligible. Derive, using the remainder of the equation, an expression for $D(r)$ in the inertial subrange, assuming that S is a constant. Compare the result with that of Problem 3 in Sect. 9.5.

10. Consider the correlation between the velocity $u_i'(\underline{x})$ and the pressure $p'(\underline{x}+\underline{r})$ in isotropic turbulence. Show that this correlation equals zero when the orientation of $u_i'(\underline{x})$ is normal to the separation $\underline{r}$. Based on this result, argue that for an arbitrary orientation of $u'(\underline{x})$, this correlation can be written as:

$$R_i(\underline{r}) = \overline{u'(\underline{x})p'(\underline{x}+\underline{r})} = f_p(r)\frac{r_i}{r},$$

with: $r = |\underline{r}|$, and where $f_p(r)$ is the correlation between $u_i'(\underline{x})$ and $p'(\underline{x}+\underline{r})$, and $u_i'(\underline{r})$ is parallel to $\underline{r}$. Next, apply the continuity equation, and show that $f_p(r)$ should satisfy:

$$\frac{\partial f_p}{\partial r} + 2\frac{f_p(r)}{r} = 0.$$

Prove, using the solution of this equation and its boundary conditions, that: $f_p(r) \equiv 0$. Link this to the *Rotta hypothesis* mentioned in Chaps. 7 and 8.

11. The equivalent to the Von Kármán–Howarth equation in the spectral domain reads

$$\frac{\partial E}{\partial t} + \frac{\partial T}{\partial \kappa} = -2\nu\kappa^2 E(\kappa),$$

where $T(\kappa)$ is here the *spectral energy transfer function*. Prove that:

$$\int_0^\infty \frac{\partial T}{\partial \kappa}\, d\kappa = T(\infty) - T(0) = 0.$$

Argue that:

(a) for small values of κ: $\partial E/\partial t = \partial T/\partial \kappa$,
(b) in the inertial subrange: $T(\kappa) = \epsilon =$ constant,
(c) for large values of κ: $\partial T/\partial \kappa = -2\nu\kappa^2 E(\kappa)$.

Sketch the shape of $T(\kappa)$ as a function of κ.

12. Consider the spectral equation from the previous problem. Suppose that in the inertial subrange, $T(\kappa) \sim E(\kappa)$. Then it follows, based on dimensional analysis, that:

$$T(\kappa) = \beta^{-1}\epsilon^{1/3}\kappa^{5/3}E(\kappa),$$

where β is the *Kolmogorov constant*. Apply this *closure hypothesis* for $T(\kappa)$ for $k \to \infty$, given that:

$$\frac{\partial T}{\partial \kappa} = -2\nu\kappa^2 E(\kappa).$$

Use this result to derive the following expression for the spectrum in the viscous region:

$$E(\kappa) = \beta\epsilon^{2/3}\kappa^{-5/3}e^{-\frac{3}{2}\beta(\kappa\eta)^{4/3}},$$

where η is the Kolmogorov length scale. This relation is known as *Pao's spectrum*; in Fig. 9.14 it is compared to some experimental data.

13. Consider the spectral shape of the Von Kármán–Howarth equation, which was introduced in Problem 11. We limit ourselves to large values of the wavenumber κ where the turbulence is isotropic, that is at least to in the inertial subrange.

(a) Show that, in this region, the following expression is approximately valid:

$$T(\kappa) + 2\nu\int_0^\kappa \kappa'^2 E(\kappa')\, d\kappa' = \epsilon.$$

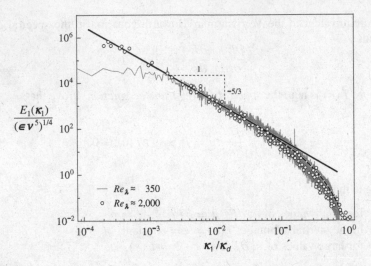

Fig. 9.14 The one-dimensional energy spectrum E_1 (defined in Problem 1 of this section) in the viscous region. The *broken line* is the expression according Pao, given in Problem 12. Experimental data were obtained in a tidal current with $Re_\lambda \approx 2,000$ (Hinze 1975) and in a turbulent jet with $Re_\lambda \approx 350$. (The Reynolds number Re_λ is based on the Taylor length scale λ and the turbulent velocity fluctuation level.) Jet data courtesy of D. Fiscaletti

(b) *Heisenberg* proposed the following *closure hypothesis* for $T(\kappa)$:

$$T(\kappa) = 2K(\kappa) \int\limits_0^\kappa \kappa'^2 E(\kappa')\, d\kappa'.$$

This relation is equivalent to *Smagorinsky's closure model* in (7.18), where the term $\int_0^\kappa \kappa'^2 E(\kappa')\, d\kappa'$ represents the deformation due to the scales at wave numbers *smaller* than κ, and $K(\kappa)$ represents the effective diffusion due to the scales at wave numbers *larger* than κ. For $K(\kappa)$ the following relation follows on the basis of a dimensional analysis:

$$K(\kappa) = \gamma \int\limits_\kappa^\infty \sqrt{\frac{E(\kappa')}{\kappa'^3}}\, d\kappa'.$$

Now substitute:

$$H(\kappa) = \int\limits_0^\kappa \kappa'^2 E(\kappa')\, d\kappa',$$

and then solve the resulting equation for H, to arrive at the result:

$$E(\kappa) = \frac{4}{3} \frac{\epsilon^{2/3}}{(2\gamma^3)^{1/3}} \frac{\kappa^{-5/3}}{(1 + C\kappa^4)},$$

where C represents an integration constant that follows from: $T(\kappa_m) = 0$ when $H = \epsilon/(2\nu)$. For $\kappa \to 0$ this spectrum satisfies the $-5/3$ law.

14. Use (9.38) and Pao's spectrum in Problem 12 above to show that the maximum in the dissipation spectrum occurs for $\kappa\eta \cong 0.2$ (for given $\beta = 1.6$).

Chapter 10
Turbulent Diffusion

As our final topic, we consider turbulent dispersion of particles and scalars (such as temperature or the mass of a specific material). We know from experience that turbulence is diffusive, and we daily use this property when we mix two fluids by stirring. This property of turbulence is of great practical importance; imagine the turbulent dispersion processes in the atmosphere. Without these, impermissibly high concentration values of pollution would occur. Now, turbulent diffusion can be described by either of two approaches, which are discussed in the following sections.

10.1   Statistical Approach

The starting point of the statistical approach of turbulent diffusion is a stationary and homogeneous turbulent velocity field. We only take into account the dispersion along one coordinate direction, for which we select the x-axis. With respect to this coordinate system we have:

$$\bar{u} = 0, \quad \text{and:} \quad \overline{u'^2} \neq f(x, t).$$

At $t = 0$ we release a labeled particle at the origin of the coordinate system, and we track the motion of this particle in the flow. We consider a so-called *material particle* that exactly follows the fluid motion. Such a particle is called *passive* (as opposed to an *inertial particle* that does not exactly follow the fluid motion and also modifies the flow). In short, the velocity of our particle is a Lagrangian variable: $u'_L(t)$, and by definition it holds that:

$$u'_L(t) = \frac{dX}{dt}, \tag{10.1}$$

where $X(t)$ is the *trajectory* of the particle along the x-axis as a function of time, with an initial condition: $X(0) = 0$.

© Springer International Publishing Switzerland 2016

F.T.M. Nieuwstadt et al., *Turbulence*, DOI 10.1007/978-3-319-31599-7_10

Fig. 10.1 Dispersion for an
ensemble of material
particles

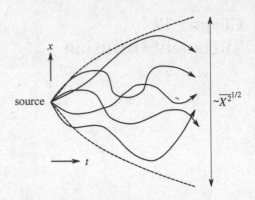

We repeat this experiment many times, or, in other words, we construct an *ensemble* of individual realizations. This can be achieved for example by releasing many particles in succession at the origin. The *dispersion* of these particles can be studied using the statistics of this ensemble. We focus here on two statistics: $\overline{X}$ and $\overline{X^2}$, which stand for the center of gravity and the cross-section, respectively, of the particle cloud that represents the ensemble, as illustrated in Fig. 10.1. (The term $\overline{X^2}$ is often referred to as the *dispersion parameter*.) We emphasize that we consider here the dispersion with respect to a fixed source (in our case the origin). We call this *absolute dispersion*.

Based on Eq. (10.1), the following equations for $\overline{X}$ and $\overline{X^2}$ can be derived:

$$\overline{X} = \int\limits_0^t \overline{u'}_L(t')\,dt', \tag{10.2}$$

$$\frac{d\frac{1}{2}\overline{X^2}}{dt} = \int\limits_0^t \overline{u'_L(t')u'_L(t)}\,dt'. \tag{10.3}$$

In the last equation we find a *Lagrangian time correlation* of the velocity at times t and t' along a trajectory of the particle, as discussed in Sect. 9.1. Due to the fact that the turbulence is stationary and homogeneous, it follows that the time correlation is only a function of the time difference $\tau = t - t'$, so that $\overline{u'_L(t')u'_L(t)} = R_L(\tau) = \overline{u'_L}^2\rho_L(\tau)$. Here, $\rho_L(\tau)$ is the *correlation coefficient*. Substitution in (10.3) and integration yields:

$$\frac{1}{2}\overline{X^2} = \overline{u'^2_L}\int\limits_0^t \left\{\int\limits_0^{t'} \rho_L(\tau)\,d\tau\right\}\,dt'. \tag{10.4}$$

When we further assume that the turbulent field is unbounded, it can be proven that: $\overline{u'}_L = \overline{u} = 0$, and that: $\overline{u'_L}^2 = \overline{u'^2}$, where $\overline{u}$ and $\overline{u'^2}$ are now *Eulerian* statistics. With this, it follows for (10.2) that:

$$\overline{X}(t) = 0. \tag{10.5}$$

Partial integration of (10.4) gives:

$$\overline{X^2}(t) = 2\,\overline{u'^2} \int_0^t (t - \tau)\rho_L(\tau)\,d\tau. \tag{10.6}$$

This equation is known as *Taylor's equation*.

As mentioned before, $\overline{X^2}(t)$ yields the cross section of the particle cloud as a function of time. We can basically only calculate this explicitly when $\rho_L(\tau)$ is given. However, we can gain further insight into the diffusion process when we use some known properties of $\rho_L(\tau)$, that is:

$$\rho_L(\tau) \approx 1 \quad \text{for: } t \to 0, \quad \text{and: } \int_0^\infty \rho_L(\tau)\,d\tau = \mathcal{T}_L \tag{10.7}$$

where $\mathcal{T}_L$ is the *Lagrangian time scale*. Using the properties in (10.7), we can derive that:

$$\overline{X^2}(t) \approx \begin{cases} \overline{u'^2}\,t^2 & \text{for: } t \to 0, \\ 2\,\overline{u'^2}\,\mathcal{T}_L\,t & \text{for: } t \to \infty. \end{cases} \tag{10.8}$$

This means that, for short times, the width of the particle cloud, which is proportional to $\sigma = (\overline{X^2})^{1/2}$, increases *linearly* with time, while at a later stage the width of the particle cloud grows at a rate proportional to $\sqrt{t}$. The latter case is called the *diffusion limit*.

Problems

1. Derive the following relation on the basis of Eq. (10.6):

$$\overline{X^2}(t) = t^2 \int_0^\infty \frac{\sin^2\left(\frac{1}{2}\omega t\right)}{\left(\frac{1}{2}\omega t\right)^2} E_L(\omega)\,d\omega,$$

where $E_L(\omega)$ is the Lagrangian spectrum that is defined as:

$$E_L(\omega) = \frac{2}{\pi} \int_0^\infty R_L(\tau)\cos(\omega\tau)\,d\tau.$$

Here we applied *Parseval's relation*, which reads:

$$\int\limits_{0}^{\infty} f(\tau)R_L(\tau)\,d\tau = \frac{\pi}{2}\int\limits_{0}^{\infty} C(\omega)E_L(\omega)\,d\omega,$$

where $C(\omega)$ is the *cosine transform* of $f(\tau)$.

Make a sketch of the factor before $E_L(\omega)$ in the equation for $\overline{X^2}$ for different times t. Use this to discuss the dispersion $\overline{X^2}(t)$ in terms of the contributions for varying eddy time scales.

2. Define the structure function $D_L(\tau) = \overline{\{u'_L(t) - u'_L(t')\}^2}$, and derive an explicit expression for it in the inertial subrange. Use this expression to determine the behavior of $\overline{X^2}(t)$ in the intermediate time domain of (10.8) when $t \sim T_L$. (See also Problem 3 of Sect. 9.5.)

3. Consider a stationary and *horizontally* homogeneous turbulent velocity field; the turbulence is non-homogeneous in the vertical (y) direction. Consider the dispersion of a fluid particle along the y-axis that is released in y_s at $t = 0$. The particle follows the fluid motion and satisfies the following equations:

$$\frac{dY}{dt} = v, \quad \frac{Dv}{Dt} = 0,$$

where we neglected viscosity.

Use a Taylor series expansion, for $t \to 0$ around y_s to show that:

$$\overline{Y} = y_s + \frac{1}{2}t^2\left.\frac{\partial\overline{v'^2}}{\partial y}\right|_{y_s} + \frac{1}{12}t^3\left.\frac{\partial^2\overline{v'^3}}{\partial y^2}\right|_{y_s}$$

$$\overline{(Y - \overline{Y})^2} = \left.\overline{v'^2}\right|_{y_s}t^2 + \frac{1}{2}t^3\left.\frac{\partial\overline{v'^3}}{\partial y}\right|_{y_s}.$$

4. Consider the *Langevin equation* as a model for the motion of a material particle along the x-axis in a homogeneous turbulent fluid with: $\overline{u} = 0$ and $\sigma_u^2 \equiv \overline{u'^2} = $ constant:

$$\frac{du}{dt} + \frac{u}{T_L} = \eta,$$

where η is a *stochastic process* with: $\overline{\eta} = 0$, and:

$$\overline{\eta(t)\eta(t')} = \frac{2\sigma_u^2}{T_L}\delta(t - t').$$

The position of the material particle is given by

$$\frac{dX}{dt} = u.$$

Solve these equations for $\overline{X'^2}$ with the following initial conditions:

(a) $\overline{u'^2}(0) = \sigma_u^2$, and: $X(0) = 0$. For these conditions, show that the particle velocity u is described by a *stationary* process. Derive that the result for $\overline{X'^2}$ agrees with the solution to Taylor's equation in (10.6) for $\rho_L(\tau) = \exp(-\tau/T_L)$.

(b) $\overline{u'^2}(0) = 0$ and $X(0) = 0$. Show for this case that:

$$\overline{X'^2} = \frac{2}{3}\sigma^2 \frac{t^3}{T_L},$$

for: $t \to 0$. This scaling is know as the *Richardson-Obukhov* scaling.

5. In a turbulent flow with a characteristic length scale $\mathcal{L}$ and a characteristic velocity scale $\mathcal{U}$, an amount of material is released. Estimate how the size of the cloud of dispersed material grows as a function of time when this size is in the scale region of the inertial subrange. (This problem is called *relative diffusion*.)
Compare the result with Problem 4b.

10.2 The Diffusion Equation

We now consider our dispersion problem from an *Eulerian* point of view, that is, in a fixed coordinate system. In the previous section we discussed the release and subsequent tracking of labelled particles. Here we consider the concentration $\chi\,[m^{-3}]$ of these particles at a fixed point as a function of time. We limit ourselves again to a passive additive, which means that χ does not influence the dynamics of our flow. The quantity χ satisfies the following conservation law:

$$\frac{D\chi}{Dt} \equiv \frac{\partial \chi}{\partial t} + u_j \frac{\partial \chi}{\partial x_j} = \mathbb{D}\frac{\partial^2 \chi}{\partial x_j^2}. \tag{10.9}$$

The left-hand side is a *material derivative* and thus describes the change of χ moving along with a fluid element. We see that this change is equal to the (molecular) *diffusion* of χ, with the diffusion coefficient $\mathbb{D}$. (The molecular diffusion of a species concentration is similar to the molecular diffusion of heat, although the physical processes are essentially different.) We neglected these molecular processes in the previous section when we discussed the statistics of the particle trajectories.

Consider a turbulent flow, for which we apply the *Reynolds decomposition*, according to:

$$\chi = \overline{\chi} + \chi', \quad \text{and:} \quad u_j = \overline{u}_j + u'_j,$$

with: $\overline{\chi'} = 0$, and: $\overline{u'} = 0$. We limit ourselves here to a stationary and homogeneous velocity field with $\overline{u}_j = 0$. Substitution in (10.9) then yields:

$$\frac{\partial \overline{\chi}}{\partial t} = -\frac{\partial}{\partial x_j} \overline{u'_j \chi'} + \mathbb{D} \frac{\partial^2 \overline{\chi}}{\partial x_j^2} \tag{10.10}$$

where $\overline{u'_j \chi'}$ represents the *turbulent concentration flux*. This equation is known as the *turbulent diffusion equation*. We can interpret the term $\overline{\chi}$ as the concentration in a certain point, which is determined as the ensemble average over a large number of point particles marked with χ. We still have to formulate the initial and boundary conditions for Eq. (10.10). As the initial condition we often take that, at $t = 0$, at a certain position, an amount Q per unit of time is introduced into the flow. This geometry is called a *point source* with *strength* Q.

The concentration flux in (10.10) is a new unknown quantity. Again we arrive at a *closure problem*. We make use of the familiar K-theory to formulate a closure hypothesis:

$$-\overline{u'_j \chi'} = K \frac{\partial \overline{\chi}}{\partial x_j}, \tag{10.11}$$

where $K \sim \mathcal{UL}$ is the turbulent diffusion coefficient, which is constant in our case of a stationary and homogeneous turbulent flow. With this, (10.10) becomes

$$\frac{\partial \overline{\chi}}{\partial t} = (K + \mathbb{D}) \frac{\partial^2 \overline{\chi}}{\partial x_j^2}. \tag{10.12}$$

The molecular diffusion coefficient $\mathbb{D}$ in gases and liquids is generally small; for example, the molecular diffusion of salt in water at $15\,^\circ\text{C}$ is $\mathbb{D} = 1.1 \times 10^{-9}$ m²/s. It is then clear that, in the equation above, the term $\mathbb{D}$ is negligible with respect to K in almost all cases, and therefore we disregard the molecular diffusion in the remainder of this chapter. The fact that $K \gg \mathbb{D}$ is of course the essence of turbulent dispersion.

Subsequently, we limit our diffusion problem to a single coordinate, for which we select the x-axis. We can achieve this in two ways: (i) We can consider a source strength Q that is uniformly distributed in the $y - z$-plane; in that case we speak of a *planar source*. (ii) We consider the integral of Eq. (10.12) over y and z, which is referred to as the *surface-integrated concentration field* that is a function of x and t only. In both cases the equation becomes:

$$\frac{\partial \overline{\chi}}{\partial t} = K \frac{\partial^2 \overline{\chi}}{\partial x^2}. \tag{10.13}$$

The solution of this equation for: $-\infty < x < \infty$, and with the initial condition: $\overline{\chi} = Q\,\delta(x)$ at $t = 0$ was already encountered in Sect. 4.1 on the Burgers equation. When we substitute the proper quantities in (4.4), we arrive at the following solution for $\overline{\chi}$:

$$\overline{\chi}(x, t) = \frac{Q}{2\sqrt{\pi K t}} e^{-\frac{x^2}{4Kt}}. \tag{10.14}$$

This solution describes a concentration distribution that spreads out in space as a function of time (see Fig. 4.1). We define the *width* σ of this concentration distribution as:

$$\sigma^2 = \int\limits_{-\infty}^{\infty} x^2\,\overline{\chi}(x, t)\,dx = 2Kt,$$

where we assume that χ is normalized with Q. This width is directly comparable with the cross-section $\overline{X^2}$ of the particle cloud that we discussed in the previous section.

When we compare the results for σ^2 and $\overline{X^2}$, we only find agreement in the so-called *diffusion limit* for $\overline{X^2}$. This means that the diffusion Eq. (10.13) can only be applied for large times, i.e. $t \gg T_L$. This can be easily understood, because for large times, the size of the cloud or the size of the concentration distribution is larger than the characteristic length scale $\mathcal{L}$ of the turbulence. This is exactly what we used in Chap. 5 as the condition for the validity of the gradient hypothesis, such as (10.11). Also, based on a comparison between σ^2 and $\overline{X^2}$, we find an explicit expression for K, that is:

$$K = \overline{u'^2}\,T_L.$$

This confirms that the turbulent exchange coefficient K is determined by the macrostructure of the turbulence.

Problems

1. Consider the wall region in a turbulent channel flow, where the turbulence is homogeneous in the x and z directions, but inhomogeneous in the y direction. We consider a uniform planar source at the wall, i.e. $y = 0$. Show that the diffusion equation for this case reads:

$$\frac{\partial \overline{\chi}}{\partial t} = \frac{\partial}{\partial y}\left(ku_*y\,\frac{\partial \overline{\chi}}{\partial y}\right),$$

with the initial condition: $\chi = Q\,\delta(y)$ at $t = 0$. Show that the solution of this problem, with the boundary condition: $\chi \to 0$ for $y \to \infty$, reads:

$$\overline{\chi} = \frac{Q}{ku_*t}e^{-\frac{y}{ku_*t}}.$$

Discuss the evolution of this result as a function of time and compare it to the solution for homogeneous turbulence.

2. Consider Eq. (2.14) for the instantaneous temperature. Apply this equation to the following initial value problem:

$$\theta = \begin{cases} 0 & \text{for: } y < 0, \\ \theta_0 & \text{for: } y > 0, \end{cases}$$

with $\theta_0 \neq f(x, z)$. The velocity field is a pure straining flow, that is: $u = \gamma x$ and $v = -\gamma y$, with $\gamma > 0$. The term γ [1/s] is the *strain rate*.

(a) Reduce the temperature Eq. (2.14) to the following simplified form for this problem:

$$\frac{\partial \theta}{\partial t} - \gamma y \frac{\partial \theta}{\partial y} = \kappa \frac{\partial^2 \theta}{\partial y^2}.$$

(b) Show that this equation has a stationary solution.
(c) Suppose that this problem describes the dynamics of the temperature fluctuations at the microscale. In that case, derive that the so-called *Corrsin scale* η_θ for temperature fluctuations, which is the equivalent of the Kolmogorov scale η for velocity fluctuations, equals:

$$\frac{\eta_\theta}{\eta} = \begin{cases} Pr^{-\frac{1}{2}} & \text{for: } Pr \geqslant 1, \\ Pr^{-\frac{3}{4}} & \text{for: } Pr \leqslant 1, \end{cases}$$

with: $Pr = \nu/\alpha$.

3. Consider a *material line segment* with a length L_0 in a turbulent flow. The line segment is deformed by the turbulence, resulting in a length *increase* (see Fig. 10.2).

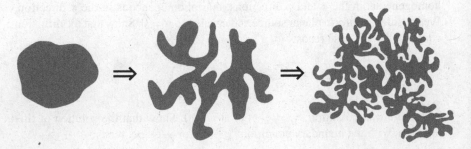

Fig. 10.2 Deformation of a material region in a turbulent flow. After: Monin and Yaglom (1973)

For this process, we introduce the following model. Assume a *cascade process*, where initially the line segment is divided into N separate elements, each with a length L_1. Subsequently, the cascade process further divides each line segment with length L_1 into N line segments, each with a length L_2. This process continues until the Kolmogorov length scale is reached.

Assume that:

$$\frac{L_1}{L_0} = \frac{L_2}{L_1} = \cdots = \beta.$$

We now introduce a *fractal dimension D*, defined as:

$$D = -\frac{\log N}{\log \beta}.$$

(a) Show that, after i steps in the cascade process, the total length L of the material line segment equals:

$$L = L_0 \left(\frac{L_i}{L_0}\right)^{1-D}.$$

(b) Now consider a *material surface* with an initial area: $S_0 \sim L_0^2$. Show that the total surface S changes during the cascade as follows:

$$S = S_0 \left(\frac{L_i}{L_0}\right)^{2-D}.$$

(c) Suppose that L_0 equals the integral scale $\mathcal{L}$. Calculate the total surface area S as a function of the Reynolds number when the cascade process has reached the Kolmogorov scale.

(d) Assume that the material surface encloses a region with concentration χ_0. This concentration can only mix with its environment due to molecular diffusion, with a diffusion coefficient $\mathbb{D}$ across the surface S. Suppose that this diffusion problem can be described by the stationary solution of (see Problem 2):

$$\frac{\partial \chi}{\partial t} - \gamma n \frac{\partial \chi}{\partial n} = \mathbb{D} \frac{\partial^2 \chi}{\partial n^2},$$

where γ [1/s] represents the strain rate of the surface, and n the coordinate normal to the surface.

Use this solution to calculate the total flux F through the surface:

$$F = \mathbb{D} \frac{\partial \overline{\chi}}{\partial n} S.$$

Show that this flux becomes independent of the Reynolds number when $D = 7/3$ (cf. Fig. 6.17). Interpret this result.

10.3   Inertial Transport

So far we have considered the dispersion of *ideal* massless particles that perfectly
follow the fluid motion, do not alter the flow, and that do not interact with each
other. In many practical situations, such as sediment transport by air or water, or the
dispersion of small droplets in a cloud, the particulate matter has a finite size and
a density that does not match the density of the fluid. This implies that the finite
inertia of the particles interacts with the fluid and affects the flow properties. This
determines the spatial and temporal distribution of the particles in the flow (referred
to as *one-way coupling*), while the inhomogeneous distribution of particles in the
fluid and the generation of wakes by the particles alter the flow properties (*two-way
coupling*), and at high concentrations the fluid motion even affects the collisions of
the particles, which in turn affects the fluid motion (*four-way coupling*).

The forces on a small rigid sphere with mass m_p and velocity $\mathbf{v}_p$ in a nonuniform
flow are given by Maxey and Riley (1983), Mei (1996), Adrian and Westerweel
(2011):

$$m_p \frac{d\mathbf{v}_p}{dt} = \mathbf{F}_{\text{G-B}} + \mathbf{F}_{\text{QS}} + \mathbf{F}_{\text{H}} + \mathbf{F}_{\text{AM}} + \mathbf{F}_{\text{FS}} + \mathbf{F}_{\text{L}}, \tag{10.15}$$

where:

$$\mathbf{F}_{\text{G-B}} = \frac{\pi}{6} d_p^3 (\rho_p - \rho_f) \mathbf{g} \tag{10.16}$$

is the weight of the particle minus the buoyancy force;

$$\mathbf{F}_{\text{QS}} = -3\pi \mu_f d_p \phi (\mathbf{v}_p - \mathbf{u}) \tag{10.17}$$

is the quasi-Stokes drag modified by a factor ϕ to account for finite Reynolds number
effects;

$$\mathbf{F}_{\text{H}} = 3\pi \mu_f \int_{-\infty}^{t} K(t - \tau) \frac{d(\mathbf{v}_p - \mathbf{u})}{dt} \, d\tau \tag{10.18}$$

is the *Basset history term* that is associated with the retarded variation of the boundary
layer around the particle (represented by the history kernel $K(t)$) as the velocity
difference between the particle and the 'free stream' outside the boundary layer
changes over time;

$$\mathbf{F}_{\text{AM}} = \frac{m_p \rho_f}{2\rho_p} \left(\frac{D\mathbf{u}}{Dt} - \frac{d\mathbf{v}_p}{dt} \right) \tag{10.19}$$

is the added mass according to the formulation of Auton et al. (1988);

$$\mathbf{F}_{\text{FS}} = m_f \frac{D\mathbf{u}}{Dt} \tag{10.20}$$

is the net force on the particle due to the variation of fluid stress around the particle; and $\mathbf{F}_L$ is a transverse lift force due to shear and particle rotation, as described by Saffman (1965), Auton (1987), McLaughlin (1993), Joseph and Ocando (2002), among others. Note that in these expressions $\frac{d}{dt}$ is the time derivative as seen by an observer moving with the particle, and $\frac{D}{Dt}$ is the material derivative of an observer moving with the fluid.

It is common to simplify the expression in (10.15) for the case of a small and heavy solid particle, that is a particle with a diameter d_p less than the Kolmogorov scale and with a density that is much larger than that of the surrounding fluid (i.e. $\rho_p \gg \rho_f$); for heavy particles, the history term (10.18) and the added mass term (10.19) become negligible, while the lift force becomes negligible for very small particles. In that case we retain for the force $\mathbf{F}$ $(=m_p d\mathbf{v}_p/dt)$ on a particle:

$$\mathbf{F} = m_p \frac{\mathbf{u} - \mathbf{v}_p}{\tau_p} + m_p \mathbf{g}, \tag{10.21}$$

where τ_p is the *particle response time*, given by:

$$\tau_p = \frac{\rho_p d_p^2}{18 \rho_f \nu}. \tag{10.22}$$

The assumption of a 'small heavy particle' means that the particle Reynolds number is very small and that the particle can be represented as a 'point particle' with finite inertia and drag force. Note that $\tau_p g \equiv w_s$ is the *terminal velocity* of a single small particle in a quiescent fluid.

The *Stokes number St* is defined as the ratio of the (Stokes) particle response time and the turbulent time scale:

$$St = \frac{\tau_p}{\tau_K}, \tag{10.23}$$

where τ_K is the Kolmogorov time scale. For $St \ll 1$ the particles accurately follow the fluid motion and thus behave as material particles (or *flow tracers*), which were described in the previous sections; for $St \gg 1$ the motion of the particles is barely affected by the flow (see also Problem 2 below). Evidently, it is the intermediate range with $St = \mathcal{O}(1)$ where the strongest interactions between the particles and fluid flow occurs.

Consider a turbulent channel flow that carries small inertial particles with a local concentration χ. Under the influence of gravity, the particles have a (concentration-dependent) terminal velocity $w_{s,m}$. Then, based on mass-conservation equation, the ensemble-averaged vertical mass flux ϕ is given by:

$$\phi \equiv -K_{\mathbb{D}} \frac{d\overline{\chi}}{dy} = w_{s,m} \overline{\chi} (1 - \overline{\chi}), \tag{10.24}$$

where $\overline{\chi}$ is the mean particle concentration, and $K_{\mathbb{D}}$ is the *eddy particle diffusivity*. We assume that the eddy particle diffusivity is proportional to the eddy viscosity K, that is: $K_{\mathbb{D}} \approx \varphi K/\beta$, where β is the turbulent Schmidt number for the particle diffusion, and φ a number that represents the damping of the turbulence due to the presence of the particles. It can be generally assumed that $\beta = 1$ and $\varphi = 1$. Furthermore, we assume that the particle concentration is low, that is $\overline{\chi} \ll 1$, and that $w_{s,m}$ is equal to the terminal velocity w_s of a single particle. Then, (10.24) reduces to:

$$K\frac{d\overline{\chi}}{dy} = -w_s\,\overline{\chi}. \qquad (10.25)$$

A simple solution for the concentration profile of sediment in a riverbed or open channel is discussed in Problem 1 at the end of this section.

Consider the particle equation of motion (10.21). Excluding the gravity term and averaging the equation over time, gives:

$$\overline{v}_p = v_f - \tau_p \overline{\frac{dv_p}{dt}},$$

with the particle derivative:

$$\frac{d}{dt} = \frac{\partial}{\partial t} + v_{p,j}\frac{\partial}{\partial x_j}.$$

Substitution of the particle derivative in the expression for $\overline{v}_p$ and multiplication with the mean concentration then gives:

$$\phi \equiv \overline{v}_p\,\overline{\chi} = \overline{\chi}\,\overline{v}_f - \overline{\chi}\,\tau_p\left(\overline{v}_p\frac{\partial \overline{v}_p}{\partial y} + \frac{\partial \overline{v'_p}^2}{\partial y}\right). \qquad (10.26)$$

(Here we use that: $\partial v_{p,j}/\partial x_j \approx 0$ for particles with a small Stokes number.) This can be written as:

$$\phi = \phi_{\text{diff}} + \phi_{\text{adv}} + \phi_{\text{turbo}},$$

where ϕ_{diff} is called the *diffusive flux* (Simonin et al. 1993), ϕ_{adv} the contribution due to the advection term in (10.21), and ϕ_{turbo} is called the *turbophoretic flux*. This last term describes the drift of particles towards regions with a low turbulence intensity, or *turbophoresis* (Caporaloni et al. 1975; Reeks 1983). This process is responsible for the uneven distribution of inertial particles in a turbulent flow.

In Fig. 10.3 are shown the particle fluxes, defined in (10.3), for a DNS of a turbulent channel flow (in the absence of gravity) with a no-slip boundary at $y^+ = 0$ and a free-slip boundary at $y^+ = 180$; see also the channel configuration in Fig. 10.5. Details of the DNS are given by van Haarlem et al. (1998). The gravity was neglected in this study to concentrate on the turbulent fluxes. Two particle types were simulated,

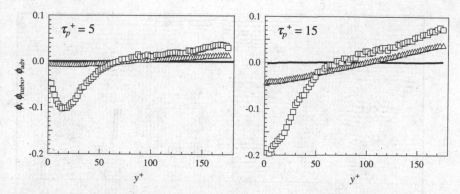

Fig. 10.3 Cross-channel particle fluxes, as defined in (10.26), for a particle-laden channel flow with a no-slip wall (at $y^+ = 0$) and a free-slip wall at $y^+ = 180$ (see Fig. 10.5) for particles with $\tau_p^+ = 5$ and 15. Plotted are: total flux ϕ ($\triangle$); turbophoretic flux ϕ_{turbo} ($\square$); and, the advection term ϕ_{adv} (——). Data obtained from a DNS of a channel flow by van Haarlem et al. (1998)

with a particle response time of τ_p^+ ($=\tau_p u_*/\nu$) of 5 and 15, respectively. A clear difference between the two particles can be observed for the turbophoretic flux near the no-slip boundary (at $y^+ = 0$). The flux $\tau_p^+ = 5$ particles increases towards the wall, reaching a maximum at $y^+ = 10$–20, while for the $\tau_p^+ = 15$ particles ϕ_{turbo} maintains to increase towards the wall. The difference can be attributed to the difference in inertia of the two particles, leading to significantly different particle deposition rates. Please note that the turbophoretic flux should be primarily balanced by the diffusive flux $\overline{v}_f \overline{\chi}$ (which could not be evaluated directly in this DNS), since the advection term ϕ_{adv} remains small over the height of the channel; see also van Haarlem et al. (1998).

The dominant flux term (in the absence of gravity) is determined by turbophoresis. This also visible in the instantaneous flow fields, where the inertial particles tend to accumulate in regions with low turbulence levels; this type of *preferential concentration* is illustrated in Fig. 10.4 for particles with $\tau_p^+ = 15$.

Problems

1. Consider a sloping riverbed or open channel with depth H that contains a particle-laden turbulent flow, as depicted in Fig. 10.5. The eddy viscosity is supposed to have the following approximate form:

$$K = \begin{cases} \dfrac{y}{H}\left(1 - \dfrac{y}{H}\right) ku_*H & \text{for: } y/H < \tfrac{1}{2} \\ \tfrac{1}{4}ku_*H & \text{for: } y/H \geqslant \tfrac{1}{2} \end{cases}$$

where k is the Von-Kármán constant and u_* the wall friction velocity. Show, using (10.25) that the mean concentration profile in the channel with respect to a reference concentration χ_a at a height $y = a$ is given by:

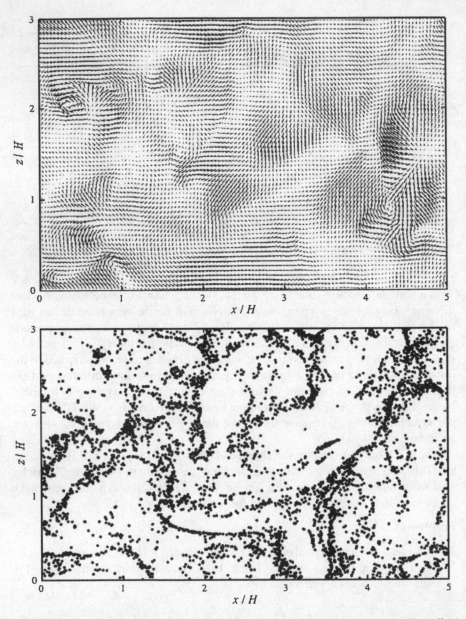

Fig. 10.4 Instantaneous results of a DNS of a particle-laden channel flow with a no-slip wall at $y^+ = 0$ and a free-slip wall at $y^+ = 180$; see Fig. 10.5. (*top*) Instantaneous flow field in the plane $y^+ = 175$ with the mean flow subtracted. (*bottom*) Instantaneous particle distribution for $\tau_p^+ = 15$ particles that lie in the domain $167 < y^+ < 175$, corresponding to the flow field above. DNS data of van Haarlem et al. (1998)

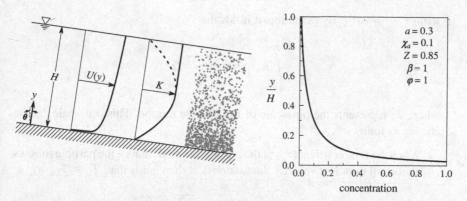

Fig. 10.5 Turbulent flow in a sloping open channel with depth H that carries sediment, with a velocity profile $U(y)$ and an eddy viscosity K. The graph on the *right* shows the concentration profile for a Rouse parameter $Z = 0.85$; see Problem 1

$$\overline{\chi} = \chi_a \cdot \begin{cases} \left[\dfrac{a}{y}\dfrac{H-y}{H-a}\right]^Z & \text{for: } y/H < \tfrac{1}{2} \\[2ex] \left[\dfrac{a}{H-a}\right]^Z \exp\left[-4Z\left(\dfrac{y}{H}-\dfrac{1}{2}\right)\right] & \text{for: } y/H \geqslant \tfrac{1}{2} \end{cases}$$

with the *Rouse parameter*:

$$Z = \frac{w_s}{ku_*}.$$

The profile for the sediment concentration, or *Rouse profile*, is show in Fig. 10.5 for $Z = 0.85$. Consider the sediment concentration profiles for $Z \gg 1$ and $Z \ll 1$.

2. Consider a small particle that is falling through a homogeneous and stationary turbulent velocity field. The density of the particle is higher than the density of the liquid. The velocity of the particle is:

$$\overline{v}_p = g\tau_p,$$

where g is the gravitational acceleration, and τ_p the *response time* of the particle. The velocity fluctuation of the particle, v'_p, compared to $\overline{v}_p$, can approximately be described by:

$$\tau_p\frac{dv'_p}{dt} + v'_p = u',$$

where u' represents the velocity fluctuation that the particle encounters on its way through the liquid. The statistical properties of u' are described by:

$$\overline{u'} = 0, \quad \text{and: } \overline{u'(t_1)u'(t_2)} = R_u(\tau),$$

with: $\tau = |t_1 - t_2|$. By definition, it holds that:

$$\int_0^\infty R_u(\tau)\,d\tau = \overline{u'^2}\, T_u,$$

where T_u represents the timescale of the particle motion. This timescale can be chosen as follows:

- When the time constant (τ_p) of the particle is small (that is, the particle follows almost all turbulent velocity fluctuations), it then holds that: $T_u \approx T_L$, or:

$$R_u(\tau) \equiv R_L(\tau) = \overline{u'^2} e^{-\tau/T_L}.$$

In other words, the particle approximates a material fluid particle.

- However, when τ_p increases, this implies that the particle can no longer follow all fluid motion. This means that the correlation function R_u has to be corrected with respect to the correlation function R_L for a material fluid element. For this we use the following model: Under the influence of gravity, the particle, so to speak, 'falls' through the eddies at an average velocity $\overline{v}_p$; The correlation function R_u can be corrected for this as follows:

$$R_u(\tau) = \overline{u'^2} e^{-\tau/T_L}\, e^{-\overline{v}_p \tau/\mathcal{L}},$$

where $\mathcal{L}$ represents here the Eulerian length scale along the trajectory of the particle. We see that in the limit $\overline{v}_p \to \infty$ (that is, a very heavy particle), the statistical properties of u' are completely described by $\mathcal{L}$ and no longer by T_L, so that:

$$R_u(\tau) = \overline{u'^2}\, e^{-\overline{v}_p \tau/\mathcal{L}}.$$

This effect is called *streamline crossing*.

The following questions arise:

(a) We assume that both u' and v'_p can be described by a *stationary process*. Show that in this case, for the correlation function of the particle velocities, $R_p = \overline{v'_p(t_1)v'_p(t_2)}$, it can be written that:

$$-\tau_p^2 \frac{d^2 R_p}{d\tau^2} + R_p(\tau) = R_u(\tau).$$

(b) By definition, it holds for R_p that:

$$\int_0^\infty R_p(\tau)\,d\tau = \overline{v'^2_p}\, T_p.$$

Use this expression to prove that:

$$\tau_p \overline{v'_p{}^2} = \overline{u'^2} T_u,$$

where τ_p and $\overline{v'_p{}^2}$ are the integral timescale and the velocity variance of the particle velocity, respectively. What does this relation imply when $\tau_p \approx T_L$?

(c) Solve the above equation for R_p. Use this solution to calculate $\overline{v'_p{}^2}$, τ_p, and λ_p, where the last variable represents the *Taylor microscale* of the particle motion, defined as:

$$\overline{\left(\frac{\partial v'_p}{\partial t}\right)^2} = \frac{\overline{v'_p{}^2}}{\lambda_p^2} = -\frac{d^2 R_p}{d\tau^2}\bigg|_{\tau=0}.$$

Erratum to: Turbulence

Frans T.M. Nieuwstadt, Bendiks J. Boersma and Jerry Westerweel

Erratum to:
F.T.M. Nieuwstadt et al., *Turbulence*,
DOI 10.1007/978-3-319-31599-7

The book was inadvertently published without including the common affiliation of the authors and about the authors in the front matter of the book. The erratum book has been updated with all the changes.

The updated original online version for this book can be found at
DOI 10.1007/978-3-319-31599-7

F.T.M. Nieuwstadt · B.J. Boersma · J. Westerweel (✉)
J.M. Burgers Center for Fluid Mechanics, Delft University of Technology,
Delft, The Netherlands
e-mail: j.westerweel@tudelft.nl

© Springer International Publishing Switzerland 2016
F.T.M. Nieuwstadt et al., *Turbulence*, DOI 10.1007/978-3-319-31599-7_11

Appendix A
Equations of Motion

Conservation of mass:	$$\frac{\partial \rho}{\partial t} + u_j \frac{\partial \rho}{\partial x_j} + \rho \frac{\partial u_i}{\partial x_i} = 0$$
incompressible fluid (continuity):	$$\frac{\partial u_i}{\partial x_i} = 0$$
Conservation of momentum:	$$\rho \frac{\partial u_i}{\partial t} + \rho u_j \frac{\partial u_i}{\partial x_j} = \rho g_i + \frac{\partial \sigma_{ij}}{\partial x_j}$$
gravitational acceleration: stress tensor:	$$g_i = (0, 0, -g)$$ $$\sigma_{ij}$$
Newtonian fluid:	$$\sigma_{ij} = -p\delta_{ij} + 2\mu s_{ij}$$
rate-of-strain tensor:	$$s_{ij} = \frac{1}{2}\left(\frac{\partial u_i}{\partial x_j} + \frac{\partial u_j}{\partial x_i}\right)$$
Navier-Stokes equations:	$$\rho \frac{\partial u_i}{\partial t} + \rho u_j \frac{\partial u_i}{\partial x_j} = \rho g_i - \frac{\partial p}{\partial x_i} + \mu \frac{\partial^2 u_i}{\partial x_j^2}$$
Equation of state: (potential) temperature:	$$\rho = f(\theta, p)$$ $$\theta$$
Conservation of energy:	$$\frac{\partial \theta}{\partial t} + u_j \frac{\partial \theta}{\partial x_j} = \kappa \frac{\partial^2 \theta}{\partial x_j^2}$$
Vorticity equation:	$$\frac{\partial \omega_i}{\partial t} + u_j \frac{\partial \omega_i}{\partial x_j} = \omega_j \frac{\partial u_i}{\partial x_j} + \nu \frac{\partial^2 \omega_i}{\partial x_j^2}$$ $$\omega_i = \epsilon_{ijk} \frac{\partial u_k}{\partial x_j}$$
Boussinesq equations:	$$\frac{\partial u_i}{\partial x_i} = 0$$ $$\frac{\partial u_i}{\partial t} + u_j \frac{\partial u_i}{\partial x_j} = -\alpha \frac{\theta}{T_0} g_i - \frac{1}{\rho_0} \frac{\partial p}{\partial x_i} + \nu \frac{\partial^2 u_i}{\partial x_j^2}$$ $$\frac{\partial \theta}{\partial t} + u_j \frac{\partial \theta}{\partial x_j} = \kappa \frac{\partial^2 \theta}{\partial x_j^2}.$$

© Springer International Publishing Switzerland 2016

F.T.M. Nieuwstadt et al., *Turbulence*, DOI 10.1007/978-3-319-31599-7

| **Reynolds-averaged Navier-Stokes equations:** | $$\frac{\partial \overline{u}_i}{\partial x_i} = 0$$ $$\frac{\partial \overline{u}_i}{\partial t} + \frac{\partial \overline{u}_i \overline{u}_j}{\partial x_j} = -\frac{1}{\rho_0}\frac{\partial \overline{p}}{\partial x_i} + \frac{g}{T_0}\overline{\theta}\delta_{i3} + \nu \frac{\partial^2 \overline{u}_i}{\partial x_i^2} - \frac{\partial \overline{u_i' u_j'}}{\partial x_j}$$ $$\frac{\partial \overline{\theta}}{\partial t} + \overline{u}_j \frac{\partial \overline{\theta}}{\partial x_j} = \kappa \frac{\partial^2 \overline{\theta}}{\partial x_i^2} - \frac{\partial \overline{u_j' \theta'}}{\partial x_j}$$ |
| **Reynolds averaging:** | $$u_i = \overline{u}_i + u_i' \quad \text{with:} \quad \overline{u_i'} = 0$$ $$p = \overline{p} + p' \quad \text{with:} \quad \overline{p'} = 0$$ $$\theta = \overline{\theta} + \theta' \quad \text{with:} \quad \overline{\theta'} = 0$$ |

Turbulent kinetic energy:
$$e = \frac{1}{2}\left(\overline{u'^2} + \overline{v'^2} + \overline{w'^2}\right) = \frac{1}{2}\overline{u_i' u_i'}$$
$$e' = \frac{1}{2}\left(u'^2 + v'^2 + w'^2\right)$$

$$\frac{De}{Dt} \equiv \frac{\partial e}{\partial t} + \overline{u}_j \frac{\partial e}{\partial x_j} = -\overline{u_i' u_j'}\frac{\partial \overline{u}_i}{\partial x_j} + \frac{g}{T_0}\overline{u_3' \theta'} - \frac{\partial}{\partial x_j}\left(\overline{u_j' e'} + \frac{1}{\rho_0}\overline{p' u_j'} + \nu \frac{\partial e}{\partial x_j}\right) - \nu \overline{\left(\frac{\partial u_i'}{\partial x_j}\right)^2}$$

per component (e_α):

$$\frac{De_\alpha}{Dt} \equiv \frac{\partial e_\alpha}{\partial t} + \overline{u}_j \frac{\partial e_\alpha}{\partial x_j} =$$
$$-\overline{u_\alpha' u_j'}\frac{\partial \overline{u}_\alpha}{\partial x_j} + \frac{g}{T_0}\overline{u_\alpha' \theta'}\delta_{j3} - \frac{\partial}{\partial x_j}\left(-\overline{u_j' e_\alpha'} - \frac{1}{\rho_0}\overline{p' u_\alpha'}\delta_{j\alpha}\right) + \frac{1}{\rho_0}\overline{p'\frac{\partial u_\alpha'}{\partial x_\alpha}} - \nu \overline{\left(\frac{\partial u_\alpha'}{\partial x_j}\right)^2}$$

Note: no summation over the index α

Reynolds stress equation:

$$\frac{D\overline{u_i' u_j'}}{Dt} \equiv \frac{\partial \overline{u_i' u_j'}}{\partial t} + \overline{u}_k \frac{\partial \overline{u_i' u_j'}}{\partial x_k} = -\overline{u_i' u_k'}\frac{\partial \overline{u}_j}{\partial x_k} - \overline{u_j' u_k'}\frac{\partial \overline{u}_i}{\partial x_k} +$$
$$-\frac{\partial}{\partial x_k}\left(\overline{u_i' u_j' u_k'} + \frac{1}{\rho_0}\overline{p' u_j'}\delta_{ik} + \frac{1}{\rho_0}\overline{p' u_i'}\delta_{jk}\right) + \frac{1}{\rho_0}\overline{p'\left(\frac{\partial u_i'}{\partial x_j} + \frac{\partial u_j'}{\partial x_i}\right)} - 2\nu \overline{\frac{\partial u_i'}{\partial x_k}}$$

| **Temperature variance:** | $$\frac{D\frac{1}{2}\overline{\theta'^2}}{Dt} \equiv \frac{\partial \frac{1}{2}\overline{\theta'^2}}{\partial t} + \overline{u}_k \frac{\partial \frac{1}{2}\overline{\theta'^2}}{\partial x_k} =$$ $$-\overline{u_j' \theta'}\frac{\partial \overline{\theta}}{\partial x_j} - \frac{\partial}{\partial x_j}\left(\overline{u_j' \frac{1}{2}\theta'^2}\right) - \kappa \overline{\left(\frac{\partial \theta'}{\partial x_j}\right)^2}$$ |

Temperature flux:

$$\frac{D\overline{u_k' \theta'}}{Dt} \equiv \frac{\partial \overline{u_k' \theta'}}{\partial t} + \overline{u}_k \frac{\partial \overline{u_k' \theta'}}{\partial x_k} = -\overline{u_j' u_k'}\frac{\partial \overline{\theta}}{\partial x_j} - \overline{u_j' \theta'}\frac{\partial \overline{u}_k}{\partial x_j} + \frac{g}{T_0}\overline{\theta'^2}\delta_{3k} - \frac{\partial}{\partial x_j}\left(\overline{u_k' u_j' \theta'}\right) + \frac{1}{\rho_0}\overline{\theta'\frac{\partial p'}{\partial x_k}}$$

Appendix B
Special Topics

B.1 Monin–Obukhov Similarity

We saw in Sect. 7.5 that buoyancy and density effects can directly affect turbulence. It seems obvious that they can, for example, also influence the average profiles of velocity and temperature. We discuss this on the basis of horizontally homogeneous turbulence close to a solid wall with an average velocity $\overline{u}_i = (\overline{u}(z), 0, 0)$. This flow configuration is discussed extensively in Chap. 6. An important application for this is the flow in the lower part of the atmosphere, which is called the *atmospheric boundary layer*; see Sect. 7.6.

Here we limit ourselves to the *wall region*. In the atmosphere this region is known as the *surface layer*. In Sect. 6.2 we argued that in this region the shear stress $\tau_t = -\overline{u'w'}$ is approximately constant and equal to

$$-\overline{u'w'} \approx u_*^2, \tag{B.1}$$

where the friction velocity u_* is defined in (6.6).

We expand this turbulent wall flow with a *temperature flux* $\overline{w'\theta'}$ (for example caused by heating or cooling of the wall). This situation is representative for the flow in our atmosphere, where the surface of the earth heats the air above it during the day (i.e., $\overline{w'\theta'} > 0$) and cools it down at night (i.e., $\overline{w'\theta'} < 0$). Analogous to (B.1), it follows for the wall region that:

$$\overline{w'\theta'} \approx \overline{w'\theta'}_s, \tag{B.2}$$

where $\overline{w'\theta'}_s$ is the temperature flux at the surface. This temperature flux $\overline{w'\theta'}_s$ adds a new characteristic parameter to our problem, but instead of directly using $\overline{w'\theta'}_s$, we consider a derived quantity. For this we use the kinetic energy equation (7.24). We neglect the time derivative and the transport terms in this equation, which is justified in the wall region. It then follows that:

© Springer International Publishing Switzerland 2016
F.T.M. Nieuwstadt et al., *Turbulence*, DOI 10.1007/978-3-319-31599-7

$$0 = u_*^2 \frac{\partial \overline{u}}{\partial z} + \frac{g}{T_0} \overline{w'\theta'}_s - \epsilon.$$

In order to calculate the velocity gradient, we consider the logarithmic velocity profile:

$$\overline{u} = \frac{u_*}{k} \ln\left(\frac{z}{z_0}\right),$$

where k is the Von-Kármán constant, and z_0 the roughness length. (The term z_0 is equivalent to the term y_0, which was introduced in Sect. 6.4 as the roughness length.) It then follows that:

$$0 = u_*^2 \frac{u_*}{kz}\left(1 - \frac{z}{L}\right) - \epsilon.$$

Here L is the so-called *Obukhov length*, which is defined as:

$$L = -\frac{u_*^3}{k\dfrac{g}{T_0}\overline{w'\theta'}_s}. \tag{B.3}$$

For $L < 0$ the flow is *unstable*, for $L > 0$ the flow is *stable*, and for $L = 0$ the flow *neutral*. This has the same consequences for the flow as discussed in Sect. 7.5. On the basis of the expressions above it follows that:

$z < L$: *shear production* is dominant;
$z > L$: *buoyant production* is dominant.

In other words, in wall turbulence the flow close to the wall is, in good approximation, always *neutral*. Density effects only come into play away from the wall.

Let us consider the Obukhov length as one of characteristic scales close to wall the wall and also consider the previously introduced scales u_*, z and z_0. Furthermore, we define a temperature scale:

$$\theta_* = -\frac{\overline{w'\theta'}_s}{u_*}. \tag{B.4}$$

We limit ourselves to the matching region or overlap region. In Problem 2 of Sect. 6.3 we argued that in this region the average gradients can be scaled with u_* and z, and we argued that z_0 does not appear in the result. In the present case, we expand these scaling parameters with L. It then follows on the basis of dimensional analysis for the dimensionless velocity gradient and the dimensionless temperature gradient in the overlap region that:

$$\frac{kz}{u_*} \frac{\partial \overline{u}}{\partial z} = \phi_m\left(\frac{z}{L}\right), \tag{B.5}$$

$$\frac{kz}{\theta_*} \frac{\partial \overline{\theta}}{\partial z} = \phi_h\left(\frac{z}{L}\right), \tag{B.6}$$

Fig. B.1 Measurements of the dimensionless velocity (ϕ_m) and temperature (ϕ_h) gradients as a function of height z from the surface relative to the Obukhov length L, defined in (B.3). The *solid lines* represent the fitted models in (B.7–B.8). After: Fleagle and Businger (1980)

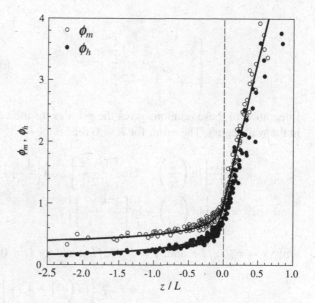

where ϕ_m and ϕ_h are two unknown functions. The expression for ϕ_m is normalized, so that for $L \to \infty$ it follows that $\phi_m(0) = 1$. Indeed, for $z/L \to 0$, the flow is *neutral*, and we retrieve the logarithmic velocity profile. The results (B.5) and (B.6) are referred to as the *Monin–Obukhov similarity*.

First we convince ourselves that these similarity relations provide adequately descriptions of the actual gradients. For this we plot in Fig. B.1 experimental data of the dimensionless velocity gradient (ϕ_m) and of the dimensionless temperature gradient (ϕ_h) as a function of $\zeta = z/L$. We find that all experimental data indeed coincide on a single curve, although it needs to be emphasized that these measurements were done at different heights. This result can be considered as experimental support for of the validity of the Monin–Obukhov similarity. It needs to be emphasized that the functions ϕ_m and ϕ_h can only be determined empirically; no arithmetical expression can be found using dimensional analysis.

By fitting a function to the experimental data from Fig. B.1 we can formulate an explicit relationship for ϕ_m and ϕ_h as a function of z/L. Although many different functional relations have been proposed, we consider here:

$$\begin{cases} \phi_m = \left(1 - 16\dfrac{z}{L}\right)^{-1/4} \\ \phi_h = \left(1 - 14\dfrac{z}{L}\right)^{-1/2} \end{cases} \qquad \text{for: } L < 0, \qquad \text{(B.7)}$$

and:

$$
\begin{cases}
\phi_m = 1 + 4.7\dfrac{z}{L} \\[2mm]
\phi_h = 1 + 4.7\dfrac{z}{L}
\end{cases}
\qquad \text{for: } L > 0. \tag{B.8}
$$

Integration of these relations gives the profiles for the velocity and the temperature in the wall region. The result for $L < 0$ reads:

$$
\overline{u} = \frac{u_*}{k}\left\{ \ln\left(\frac{z}{z_0}\right) - 2\ln\left[\frac{1+x}{2}\right] - \ln\left[\frac{1+x^2}{2}\right] + 2\arctan x - \frac{\pi}{2} \right\},
$$
$$
\overline{\theta} - T_s = \frac{\theta_*}{k}\left\{ \ln\left(\frac{z}{z_0}\right) - \ln\left[\frac{1+y^2}{2}\right] \right\}, \tag{B.9}
$$

with: $x = \phi_m^{-1}(z/L)$, and: $y = \phi_h^{-1}(z/L)$, and for $L > 0$:

$$
\overline{u} = \frac{u_*}{k}\left\{ \ln\left(\frac{z}{z_0}\right) + 4.7\frac{z}{L} \right\},
$$
$$
\overline{\theta} - T_s = \frac{\theta_*}{k}\left\{ \ln\left(\frac{z}{z_0}\right) + 4.7\frac{z}{L} \right\}. \tag{B.10}
$$

Here we applied the boundary condition: $\overline{u}(z_0) = 0$, and we introduced the *surface temperature* T_s, although formally T_s represents the temperature at the height $z = z_0$. (In applying these boundary conditions we have neglected the terms with z_0/L, because in general $z_0 \ll L$ and close to the wall the flow can be considered neutral.)

The equations above for the profiles of the average velocity and the average temperature are very useful in practice. This is because they specify a relationship between the turbulent fluxes: $-\overline{u'w'} \equiv u_*^2$ and $\overline{w'\theta'}_s$, and the average profiles $\overline{u}$ and $\overline{\theta}$. From this it follows that we can derive the fluxes $-\overline{u'w'}$ and $\overline{w'\theta'}_s$ from observations of $\overline{u}$ and $\overline{\theta}$ at various heights without having to use any direct measurements that require expensive and complicated measurement equipment. This is referred to as the *profile method*.

It appears that the Monin–Obukhov similarity can also be applied to other turbulent quantities in the wall region. Here we demonstrate this for the vertical velocity variance $\overline{w'^2}$ and the temperature variance $\overline{\theta'^2}$. On the basis of dimensional analysis, it follows that:

$$
\overline{w'^2} = u_*^2\, f_w\left(\frac{z}{L}\right),
$$
$$
\overline{\theta'^2} = \theta_*^2\, f_\theta\left(\frac{z}{L}\right).
$$

In Fig. B.2 we show some experimental data for these quantities, which indeed confirm the validity of the Monin–Obukhov similarity for these quantities. However, other experimental data appear to show that the Monin–Obukhov similarity does not hold for the *horizontal* velocity variance $\overline{u'^2}$.

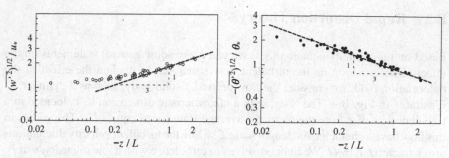

Fig. B.2 Observations of $(\overline{w'^2})^{1/2}$ and $(\overline{\theta'^2})^{1/2}$ as a function of z/L according to the Monin–Obukhov similarity. From: Wyngaard et al. (1971)

Problems

1. Using (B.7) and (B.8), calculate the exchange coefficients K and K_H as a function of z/L. Plot these relations in a single graph and show that in unstable circumstances the exchange coefficients are larger than in stable circumstances. Is this physically explainable?

2. Calculate the flux Richardson number Ri_f defined in (7.26) in the wall region. Derive that for $L > 0$ the critical value of Ri_f equals $Ri_f^{(cr)} = 5.2^{-1} = 0.19$.

3. The condition $z/L \gg 1$ is called *free convection*. In that case, turbulence production by shear is virtually negligible.

   (a) Calculate that u_* is no longer relevant as a velocity scale.
   (b) Derive that, in this case, it holds that:

   $$\overline{w'^2}^{1/2} \sim u_* \left(\tfrac{z}{L}\right)^{1/3},$$
   $$\overline{\theta'^2}^{1/2} \sim -\theta_* \left(\tfrac{z}{L}\right)^{-1/3}.$$

   Compare this result with the experimental data in Fig. B.2.
   (c) Show that in this case the eddy diffusion coefficient K_H is given by:

   $$K_H = c \left(\frac{g}{T_0} \overline{w'\theta'}_s \right)^{1/3} z^{4/3},$$

   where c is a constant. Using this expression for K_H, calculate the temperature profile $\overline{\theta}$. Does this result agree with (B.9)?

4. Can you think of a reason why the Monin–Obukhov similarity is not valid for the horizontal velocity variance $\overline{u'^2}$?

B.2   Rapid Distortion Theory

Based on the vorticity equation (8.3) we can make some general statements on the
effect of deformation on the turbulent flow structure. To explain the effect of the
deformation field, we consider the flow around a solid body, such as a cylinder, as
illustrated in Fig. B.3. The body, with a characteristic dimension d, is located in a
turbulent flow. Far from the body, the average flow speed equals U_0. The eddies in
this flow have a characteristic length scale $\mathcal{L}$, while the turbulent velocity fluctuations
are characterized by $\mathcal{U}$. We limit ourselves to turbulence with a low intensity, that is:
$\mathcal{U} \ll U_0$.

In a thin layer around the front of the body, indicated by B in Fig. B.3, shear forces
(both viscous and turbulent) are present, satisfying the boundary condition for the
tangential velocity with respect to the body surface. The thickness of this layer is δ
and when the Reynolds number, based on U_0 and d, is sufficiently large, then: $\delta \ll \mathcal{L}$.
Under these conditions the effect of this boundary layer can be neglected. Generally, a
large region with separated flow or wake, is found behind the body, indicated by W in
Fig. B.3. For the present analysis we disregard the interaction between the wake and
the turbulence outside the wake. Finally, we identify a region T, where inertial forces
dominate. These inertial forces relate to the deformation related to the presence of
the body in the flow field. The size of this region, which is thus characterized by the
deformation field, scales in a first approximation with the size of the body, d.

We now distinguish two cases: $\mathcal{L} < d$, and: $\mathcal{L} > d$.

$\mathcal{L} < d$: Deformation Dominates

In this case, the eddies are smaller than the size of the body. We can imagine that
these eddies in region T are deformed by the deformation field that relates to the
presence of the body. In a first approximation, the eddies are then transported at an

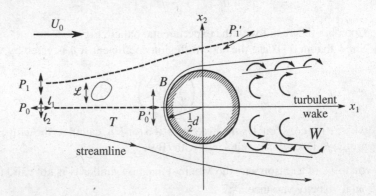

Fig. B.3 A cylinder with diameter d in a turbulent flow with a mean velocity U_0 and a typical eddy
length scale $\mathcal{L}$. Line elements are deformed when approaching $(P_0 - P_0')$ and passing along $(P_1 - P_1')$
the cylinder. B thin boundary layer in front of the cylinder; T turbulent flow region dominated by
inertial motion; W wake or separation region behind the cylinder. From: Britter et al. (1979)

average speed of order $\mathcal{O}(U_0)$ (see Sect. 9.4). The characteristic timescale T_d for the deformation of the eddies is then given by:

$$T_d \approx \frac{d}{U_0}. \tag{B.11}$$

Additionally, the characteristic timescale $\mathcal{T}$ of the eddies is defined as

$$\mathcal{T} \approx \frac{\mathcal{L}}{\mathcal{U}}. \tag{B.12}$$

We now assume that:

$$T_d \ll \mathcal{T}, \tag{B.13}$$

or the timescale of the deformation is much smaller than the characteristic timescale of the eddies. We therefore call this condition *rapid distortion*. In physical terms, this means that the turbulent vortices do not have any time, so to speak, to exchange energy during the deformation phase, for example through the cascade process. The relations (B.11–B.13) then imply:

$$\frac{\mathcal{U}}{U_0} \ll \frac{\mathcal{L}}{d}.$$

Because we already said that $\mathcal{L} < d$, it follows that $\mathcal{U} \ll U_0$. In other words, we have to limit ourselves to turbulence with a very weak intensity.

We write the instantaneous velocity and vorticity as:

$$u_i = \overline{u}_i + u_i', \quad \text{and:} \quad \omega_i = \omega_i',$$

where we adopt the notation that a prime indicates the fluctuating part. The average velocity is $\overline{u}_i$, which has a magnitude $\mathcal{O}(U_0)$. The average vorticity equals zero, which implies that the mean flow can be described as *potential flow* (Kundu and Cohen 2004). This is not essential to the theory, but is considered here for reasons of simplicity. We substitute these expressions for the velocity and vorticity in the vorticity equation (8.3) to arrive at the following equation:

$$\frac{\partial \omega_i'}{\partial t} + \overline{u}_j \frac{\partial \omega_i'}{\partial x_j} + u_j' \frac{\partial \omega_i'}{\partial x_j} - \overline{u_j' \frac{\partial \omega_i'}{\partial x_j}} = \omega_j' \frac{\partial u_i'}{\partial x_j} + \nu \frac{\partial^2 \omega_i'}{\partial x_j^2} + \omega_j' \frac{\partial \overline{u}_i}{\partial x_j} - \overline{\omega_j' \frac{\partial u_i'}{\partial x_j}}. \tag{B.14}$$

The last two terms on the left-hand side of this equation represent the advection of the vorticity fluctuations by the fluctuating velocity field. Because we assumed that the turbulence intensity is weak, i.e.,

$$u_i' \sim \mathcal{U} \ll \overline{u}_i \sim \mathcal{O}(U_0),$$

it now follows that these terms are negligible compared to the advection by the mean flow. The last two terms on the right-hand side of (B.14) represent a change of the vorticity fluctuations due to the deformation by the fluctuating velocity field. In Sect. 8.1 we argued that this term is represents the cascade process, that is the interaction between the eddies. As we stated before, the time scale $\mathcal{T}$ coupled to this interaction is in this case larger than the time scale T_d. In other words, these terms on the right-hand side of (B.14) are negligible compared to the first term that represents the deformation by the average velocity field. Finally, we notice that the Reynolds number (as applied to the macrostructure): $\mathcal{L}\mathcal{U}/\nu$ is generally much larger than 1. From this it follows that the viscous term on the right-hand side of (B.14) can be neglected. We thus limit ourselves to the dynamics of the larger and most energetic eddies.

We now obtain the following *linearized* vorticity equation:

$$\frac{\partial \omega_i'}{\partial t} + \overline{u}_j \frac{\partial \omega_i'}{\partial x_j} = \omega_j' \frac{\partial \overline{u}_i}{\partial x_j}. \tag{B.15}$$

When the average (irrotational) velocity field is given, this equation can be solved as an initial value problem. Because the equation is linear, the solution is found by means of standard methods. This means that, given the turbulence characteristics of the inflowing fluid, we can explicitly calculate the turbulence around the body.

In the remainder of this case we limit ourselves to a quantitative evaluation of this solution, where we use the concept of vortex stretching introduced in Sect. 8.1. Consider two material line segments ℓ_1 and ℓ_2 at a point P_0 far away from the cylinder at the symmetry line, as indicated in Fig. B.3). When the flow approaches the cylinder, line segment ℓ_1, which is oriented in the direction of the x_1-axis, compresses, while the line element ℓ_2 that is oriented in the direction of the x_2-axis becomes stretched. The resulting configuration is shown in point P_0'. In Problem 3 of Sect. 8.1 Cauchy's solution was deduced; this essentially says that vorticity in a non-viscous flow behaves like a material line segment. Thus, we can directly translate the behavior of the line segments in Fig. B.3 to vorticity. The vorticity along the x_2-axis thus increases due to vortex stretching, while the vorticity along the x_1-axis is reduced. Vorticity along the x_2-axis is coupled to the velocity components u_1 and u_3, while vorticity along the x_1-axis is coupled to the components u_2 and u_3. It thus follows that the u_1 component increases and that u_2 decreases when we approach the cylinder. In short, the fluctuations normal to the wall of our body *increase*; this is a rather surprising result.

We can expand our consideration of the flow to other places than the symmetry-axis. An example is given in Fig. B.3. Besides a change in length, the line element now also experiences a change in orientation. In Sect. 8.1 we described this process, where through an angular rotation of line segments, vorticity changes its orientation. This means that a vortex line, which was oriented in the direction of the x_2-axis at P_1, obtains a component in the x_1-direction at P_1'. The result is that a velocity component in the x_2-direction is induced, while at first (in P_1) only the velocity components u_1 and u_3 had non-zero values. The orientation of the vortex line segment at P_1' suggest that the components u_1 and u_2 are correlated. In other words, there is a Reynolds stress $-\overline{u_1' u_2'}$ here, which was not yet present at P_1. Hence, Reynolds stresses can be generated by the deformation of a turbulent flow field.

$\mathcal{L} > d$: Blocking Dominates

In the preceding case ($\mathcal{L} < d$) we neglected the fact that turbulent eddies at the body should satisfy a boundary condition for the normal component of the velocity, since the fluid can not penetrate the body. This is called *blocking*. This boundary condition affects the structure of the eddy, which has a dimension $\mathcal{O}(\mathcal{L})$. We stated above that the length scale over which the deformation field acts is limited to an area with a dimension $\mathcal{O}(d)$. So, whenever $\mathcal{L} > d$, the flow is dominated by blocking.

The velocity field can now be written as:

$$u_i = u_i^{(0)} + u_i^{(b)}, \tag{B.16}$$

where $u_i^{(0)}$ is the instantaneous velocity far away from the body. The term $u_i^{(b)}$ is the velocity field induced by blocking. Because we have neglected the deformation field, $u_i^{(b)}$ can only be created by pressure forces. Pressure forces cannot produce vorticity (at least, not in a velocity field with a homogeneous density). This result is known in fluid mechanics as *Kelvins theorem* (Kundu and Cohen 2004). Therefore the velocity field has to satisfy the following equations in the case of blocking:

$$\nabla \cdot \overline{u}^{(b)} = 0, \quad \text{and:} \quad \nabla \times \underline{u}^{(b)} = 0, \tag{B.17}$$

with the boundary condition:

$$\underline{u}^{(b)} \cdot \underline{n} = -u_n^{(0)},$$

where $\underline{n}$ represents the normal to the body surface and $u_n^{(0)}$ the normal component of the undisturbed velocity. This boundary condition ensures that the normal velocity component at the body surface equals zero. Based on Eq. (B.17), it follows that, using potential flow theory, the blocking can be solved for an irrotational flow.

The behavior of the solution can be easily seen when we imagine the blocking effect. When the flow moves from P_0 to P_0' in Fig. B.3, the normal component of velocity u_1 decreases until it eventually reaches zero at the body surface. On the other hand, the tangential velocity component increases. By the blocking, energy is transferred from the wall-normal velocity component to the tangential one. Hence, we notice that in this case the behavior of the u_1-component and the u_2-component of the velocity is the opposite of what we found for the case $\mathcal{L} < d$. This has been confirmed by experimental data. An example is shown in Fig. B.4, where the component of the velocity fluctuations along the symmetry-axis of a cylinder flow is shown.

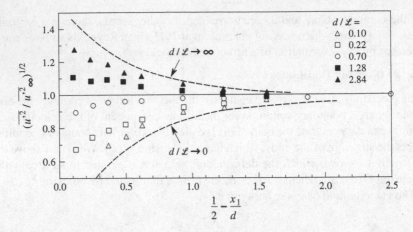

Fig. B.4 The component of the velocity fluctuations along the symmetry-axis of a turbulent flow approaching a cylinder for various values of $2\mathcal{L}/d$, where d is the cylinder diameter. The *closed symbols* are for the case $\mathcal{L} < d$ and the *open symbols* for the case $\mathcal{L} > d$. Data from: Britter et al. (1979)

B.3 Aeroacoustics

Liquids and gasses flowing around obstacles often induce periodic flow behavior. One of the well known examples of this is the Von Kármán vortex street behind a (circular) cylinder. Under certain conditions these flows can also generate audible sound. Examples are numerous, e.g., the noise generated by a high speed train, by the side mirror of a car, or by flowing water in a small river. Unbounded high speed flows also generate sound, e.g., the sound generated by high speed jets of modern aircrafts. The loudness of sound is often characterized by the so-called *sound pressure level* (SPL), which is defined as:

$$\text{SPL} = 20 \log \left(\frac{\sqrt{\overline{p'^2}}}{p_{\text{ref}}} \right) \text{ (dB)}$$

where p' is the acoustic pressure fluctuation, and p_{ref} a reference value, which for air takes the value $p_{\text{ref}} = 2 \times 10^{-5}$ Pa. A sound pressure level of 100 dB thus corresponds to a root mean square pressure value of only 1 Pa. The acoustic component of the pressure is in general several orders smaller than the hydrodynamic pressure (which scales with ρU^2), which is again much smaller than the ambient pressure.

In the early 1950s Sir James Lighthill (1952) developed a theoretical framework connecting aerodynamics and acoustics. Nowadays this field of research is known as *aeroacoustics*. Below this theory is considered shortly. A concise introduction is given in the book by Howe (2003).

Consider the equations for the conservation of mass and momentum for *compressible* flow:

$$\frac{\partial \rho}{\partial t} + \frac{\partial \rho u_i}{\partial x_i} = 0, \tag{B.18}$$

$$\frac{\partial \rho u_i}{\partial t} + \frac{\partial \rho u_i u_j}{\partial x_j} = -\frac{\partial p}{\partial x_i} + \frac{\partial}{\partial x_j}\tau_{ij}, \tag{B.19}$$

where ρ is the fluid density, u_i the fluid velocity vector, p the pressure, and τ_{ij} the viscous stress tensor. The divergence operator $\partial/\partial x_i$ is applied to the equation for the conservation of momentum, and the time derivative $\partial/\partial t$ to the equation for the conservation of mass, resulting in:

$$\frac{\partial^2 \rho}{\partial t^2} + \frac{\partial^2}{\partial t \partial x_i}\rho u_i = 0, \tag{B.20}$$

$$\frac{\partial^2}{\partial x_i \partial t}\rho u_i + \frac{\partial^2}{\partial x_i \partial x_j}\rho u_i u_j = -\frac{\partial^2 p}{\partial x_i^2} + \frac{\partial^2}{\partial x_i \partial x_j}\tau_{ij}. \tag{B.21}$$

With help of Eq. (B.20), Eq. (B.21) can be rewritten as:

$$-\frac{\partial^2 \rho}{\partial t^2} + \frac{\partial^2 p}{\partial x_i^2} = \frac{\partial^2}{\partial x_i \partial x_j}\left[-\rho u_i u_j + \tau_{ij}\right]. \tag{B.22}$$

For isentropic flow (denoted by the subscript 's') and a small (linear) acoustic disturbance, the following holds

$$c^2 = \left(\frac{\partial p}{\partial \rho}\right)_s = \frac{p - p_0}{\rho - \rho_0} = \frac{p'}{\rho'},$$

where c is the speed of sound, and the subscript '0' denotes the reference or ambient state. With the help of this relation, Eq. (B.22) can be written as:

$$\frac{\partial^2 \rho'}{\partial t^2} - c^2\frac{\partial^2 \rho'}{\partial x_i^2} = \frac{\partial^2}{\partial x_i x_j}\left[\rho u_i u_j + \delta_{ij}\left(p' - c^2\rho'\right)\tau_{ij}\right], \tag{B.23}$$

where we used the simple mathematical relation: $p' = c^2\rho' + (p' - c^2\rho')$ to convert the pressure fluctuation p' in a density fluctuation ρ'. Equation (B.23) is the well-known *Lighthill equation* for the acoustic density fluctuation ρ'. The term on the right hand side is often referred to as the *Lighthill stress tensor* T_{ij}:

$$T_{ij} \equiv \rho u_i u_j + \delta_{ij}\left(p' - c^2\rho'\right) - \tau_{ij}.$$

The Lighthill equation (B.23) is an exact reformulation of the Navier-Stokes equations. The operator on the left hand side side takes the form of a wave equation. The term on the right hand side can be seen as a forcing term. In the present form this equation cannot be solved directly. However, some simplifications can be made for flows with a high Reynolds number and low Mach number. For small Mach numbers $Ma = u/c$ it can be assumed that: $\rho' \ll \rho_o$. Furthermore, for high Reynolds number flows the viscous stresses can be ignored, and the right hand side of (B.22) reduces to: $\rho_o u_i u_j$, and Eq. (B.23) then takes the form:

$$\frac{\partial^2 \rho'}{\partial t^2} - c^2 \frac{\partial^2 \rho'}{\partial x_i^2} = \frac{\partial^2}{\partial x_i x_j} \rho_o u_i u_j. \tag{B.24}$$

From the equation above it is clear that there is a relation between non-linear flow phenomena, the term $u_i u_j$ and acoustics. The solution of this equation can be obtained via a free-space Green's function; see e.g., Howe (2003):

$$p'(x_i, t) = \frac{1}{4\pi} \frac{\partial^2}{\partial x_i x_j} \int_{-\infty}^{\infty} \frac{\rho_o u_i u_j \, (y_i, t - |x_i - y_i|/c)}{|x_i - y_i|} d^3 y_i, \tag{B.25}$$

where x_i is the position of the observer of the sound with respect to the origin on the coordinate system, and y_i the vector from the origin of the coordinate system to the source location. The term between parenthesis indicates that the term $\rho_o u_i u_j$ should be evaluated at position y_i at time $t = |x_i - y_i|/c$, i.e., in the past. For a compact source, i.e., a source with: (i) a spatial extent that is much smaller than the wavelength of the acoustic wave it generates, and (ii) with the observer far away of the acoustic source, it can be assumed that: $|y_i| \ll |x_i|$, so that the equation above can be reformulated as:

$$p'(x_i, t) = \frac{1}{4\pi} \frac{\partial^2}{\partial x_i x_j} \int_{-\infty}^{\infty} \frac{\rho_o u_i u_j \, (y_i, t - |x_i|/c)}{|x_i|} d^3 y_i. \tag{B.26}$$

Furthermore, far away from the source region, the spatial derivative can be replaced by[1]

$$\frac{\partial}{\partial x_j} = -\frac{1}{c} \frac{x_j}{|x_i|} \frac{\partial}{\partial t}.$$

Then, (B.26) can be written as:

---

[1]

$$\text{with: } \frac{\partial f(\boldsymbol{x}, t)}{\partial x_j} = \frac{\partial f(\boldsymbol{x}, t)}{\partial t} \frac{\partial}{\partial x_j} (\boldsymbol{x}, t) = \frac{\partial f(\boldsymbol{x}, t)}{\partial t} \frac{x_j}{|\boldsymbol{x}|}.$$

$$p'(x_i, t) = \frac{x_i x_j}{4\pi c^2 |\boldsymbol{x}|^3} \frac{\partial^2}{\partial t^2} \int_{-\infty}^{\infty} \frac{\rho_o u_i u_j \, (y_i, t - |x_i|/c)}{|x_i|} d^3 y_i. \tag{B.27}$$

This equation gives a prediction of the acoustic far field of a localized turbulent source region.

Acoustic Emissions of a Jet

Lighthill developed his theory primarily to predict the acoustic emissions of turbulent jets. He assumed that the non-linear term $\rho_o u_i u_j$, in Eq. (B.27) scales with a velocity scale $\mathcal{U}$, i.e., $\rho u_i u_j = \mathcal{O}(\rho_o \mathcal{U}^2)$. The non-linear term evolves with the flow, and therefore the derivative $\partial/\partial t$ of the non-linear term scales with the flow velocity $\mathcal{U}$ and length scale $\mathcal{L}$, i.e.,

$$\frac{\partial^2}{\partial t^2} \int \rho_o u_i u_j \, dy^3 = \mathcal{O}\left(\rho_o \mathcal{U}^4 \mathcal{L}\right).$$

The acoustic pressure fluctuation p' in Eq. (B.27) thus scales as:

$$p' = \mathcal{O}\left(\frac{1}{c^2} \mathcal{U}^4 \mathcal{L}\right).$$

The acoustic power emitted by an eddy at the observer location $|x_i|$ is determined by a surface integral of the acoustic intensity $I = p'^2/(\rho_o c)$ over a sphere centered at the eddy, i.e.,

$$\text{power} \approx 4\pi |x_i|^2 \frac{p'^2}{\rho_o c} = \rho_o \mathcal{L}^2 \frac{\mathcal{U}^8}{c^5}.$$

This is the famous *eighth-power law* for acoustic emission, which illustrates the effect of increasing flow speed on the acoustic power.

Lighthill's theory is very useful for understanding the relation between flow, turbulence and acoustics, but it is not very suited for actual acoustic predictions. Important effects, such as the advection of the acoustic source and the refraction of sound by the flow, are not included in the theory. Since the late 1950s other techniques have been developed that give a better prediction of acoustic emissions from a flow field. For an overview of the techniques we refer to the work of Howe (2003).

To illustrate certain features of jet acoustics, some results of numerical simulations (Moore 2009) are included. In Fig. B.5 the near and far acoustic fields of a turbulent jet are shown. The acoustic field is visualized with help of the divergence of the velocity (dilatation), which in the far field is directly related to the acoustic pressure fluctuation, via

$$\frac{\partial p'}{\partial t} = -\rho c^2 \frac{\partial u_i}{\partial x_i}.$$

Fig. B.5 The dilatation ($\partial u_i / \partial x_i$) overlaid with the vorticity amplitude field in a subsonic jet with a Reynolds number of 4,000 and a Mach number of 0.9, based on nozzle conditions. From: Moore (2009)

Dilatation waves (i.e., sound waves) are generated in regions with large spatial fluctuations in the vorticity. In Fig. B.6 is shown the acoustic far field of the jet, which is is here obtained by means of the so-called *Ffowcs-Williams and Hawkings* (FWH)

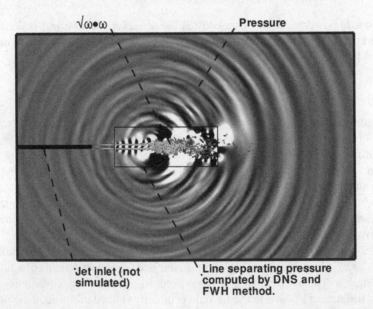

Fig. B.6 Contours of vorticity amplitude ($\sqrt{\omega_i \omega_i}$), superimposed on contours of pressure. Note that Mach and Reynolds numbers are identical to those in Fig. B.5. Differences between the flow fields in this figure and the previous figure are due to differences in the initial velocity profile. Within the *rectangular box*, the pressure field is computed by means of *direct numerical simulation* (DNS), while outside the box it is computed by means of the *Ffowcs-Williams and Hawkings* (FWH) equation (Howe 2003)

equation (Howe 2003). Sound is predominately emitted at angles of about 30°–40° with respect to the jet centerline. At smaller angles with the jet centerline the amplitude of the acoustic waves decreases. This is due to the fact that sound waves are refracted by the turbulent flow field. The area with relatively low sound levels in the direction of the jet is sometimes referred to as the *'cone of silence.'*

B.4 Rotating Turbulence

Herman Clercx

This section briefly sketches the effect of background rotation on turbulent flows, which is of relevance in many geophysical and astrophysical flows and in many industrial flows (such as in rotating machinery). Statistical homogeneity and isotropy cannot be a starting point for the basic analysis of rotating turbulence as with the presence of background rotation anisotropy is introduced. This anisotropy distinguishes the coordinate parallel to the rotation axis with those perpendicular to the rotation axis. At best we can thus assume statistical homogeneity (in 3D) and statistical isotropy in the 2D plane perpendicular to the rotation axis (provided the turbulent flow is not confined).

Depending on the integral-scale Reynolds number and the rotation rate (usually quantified by the *Rossby number*, $Ro \propto \Omega^{-1}$; see 'basic equations' below) a distinction can be made between three-dimensional (3D) homogeneous isotropic turbulence, quasi-two-dimensional (2D) turbulence and wave turbulence. In the schematic in Fig. B.7 the different regimes are illustrated in the (Re, Ro)-plane. In the linear low-Reynolds number regime in rapid rotating fluids ($Ro \ll 1$) inertial waves are observed. For similar rotation rates but increased Reynolds number non-linear interactions become dominant and we arrive in the wave-turbulence regime. Finally, for

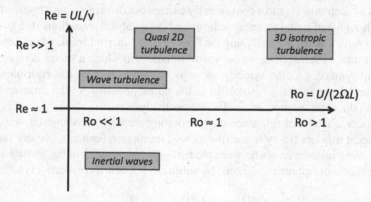

Fig. B.7 A very schematic representation of the different regimes in rotating turbulence in the (Re, Ro)-plane

rotation-dominated turbulence ($Ro < 1$) at high Reynolds number we arrive at a kind of quasi-two-dimensional turbulence in planes perpendicular to the rotation axis.

One of the famous experiments exploring the role of background rotation on turbulence was conducted by Hopfinger et al. (1982). They investigated the large-scale effects induced by rotation on turbulence. Their experiment consisted of a large container filled with fluid, rotating about its vertical axis. Near the bottom of the container the flow was continuously forced by a vertically oscillating grid (i.e., the grid motion was parallel to the rotation axis). The turbulent flow in the neighborhood of the oscillating grid, where typical turbulence time scales are much smaller than the rotation time scale ($Re \gg 1$ and $Ro > 1$), is more or less unaffected by rotation and behaves thus like 3D homogeneous isotropic turbulence. However, sufficiently far above the oscillating grid, where the turbulence intensity has already decreased such that characteristic advection time scales have become substantially larger than the rotation time scale ($Ro < 1$), vertically-aligned vortical structures have been observed (and an overall increase of the turbulent integral length scale with increasing rotation rate). At first sight one might thus conclude that the turbulent flow behaves quasi-two-dimensional, roughly independent of the vertical axis.

Which processes in rotating flows are responsible for the specific statistical properties and large-scale anisotropic flow structuring of rotating turbulence? Below we provide a concise introduction into some of these aspects including quasi-two-dimensionalization of the flow (Taylor–Proudman theorem), the role of domain boundaries and rotating boundary layer flows, and the origin of inertial waves. For a more detailed analysis of rotating flows the monograph by Greenspan (1968) can be consulted. For rotating flows in a geophysical context the reader is referred to the textbook by Pedlosky (1987). For an extensive overview of the theory of rotating and stratified turbulence the reader is referred to a recent monograph by Davidson (2013).

Basic Equations

Background rotation can strongly affect the dynamics of turbulent flows. A proper analysis of such flows is most conveniently carried out in a co-rotating frame of reference although additional terms appear in the Navier-Stokes equations (the so-called *fictitious forces* as we are analyzing the flow in a non-inertial frame of reference). In the remainder of this section we use vector notation, in which a vector is represented by a bold symbol, e.g., the velocity vector is written as $\mathbf{u}$ (which is equivalent to u_i and $\underline{u}$), and the gradient $\frac{\partial}{\partial x_i}$ is written as the *nabla operator* ∇; the inner and outer vector products are written as $\cdot$ and $\times$, respectively.

Consider a frame of reference that is rotating steadily with angular velocity $\mathbf{\Omega}$ with respect to a laboratory frame (the inertial frame of reference). For any vector $\mathbf{Q}$ the following relation exists between the time-derivative of $\mathbf{Q}$ in the inertial and the rotating frame of reference (denoted by subindices I and R, respectively):

$$\left(\frac{d\mathbf{Q}}{dt}\right)_I = \left(\frac{d\mathbf{Q}}{dt}\right)_R + \mathbf{\Omega} \times \mathbf{Q}. \qquad (B.28)$$

From this relation we obtain for the position vector $\mathbf{r}$ of a fluid element:

$$\left(\frac{d\mathbf{r}}{dt}\right)_I = \left(\frac{d\mathbf{r}}{dt}\right)_R + \mathbf{\Omega} \times \mathbf{r}, \tag{B.29}$$

or: $\mathbf{u}_I = \mathbf{u}_R + \mathbf{\Omega} \times \mathbf{r}$. Applying relation (B.28) once again with $\mathbf{u}_I$ gives the relation between the acceleration of a fluid element in the inertial frame and the rotating frame of reference

$$\left(\frac{d\mathbf{u}_I}{dt}\right)_I = \left(\frac{d\mathbf{u}_I}{dt}\right)_R + \mathbf{\Omega} \times \mathbf{u}_I = \left(\frac{d\mathbf{u}_R}{dt}\right)_R + 2\mathbf{\Omega} \times \mathbf{u}_R + \mathbf{\Omega} \times (\mathbf{\Omega} \times \mathbf{r}). \tag{B.30}$$

Here, $2\mathbf{\Omega} \times \mathbf{u}_R$ is the *Coriolis acceleration* and $\mathbf{\Omega} \times (\mathbf{\Omega} \times \mathbf{r})$ is the *centrifugal acceleration*. The latter contribution can be reformulated as a gradient of a scalar: $-\nabla(\frac{1}{2}\Omega^2 r_\perp^2)$, with $r_\perp$ the distance of the point $\mathbf{r}$ from the rotation axis and $\Omega = |\mathbf{\Omega}|$. With (B.30) the Navier-Stokes equation in the rotating frame of reference reads

$$\frac{\partial \mathbf{u}}{\partial t} + (\mathbf{u} \cdot \nabla)\mathbf{u} + 2\mathbf{\Omega} \times \mathbf{u} = -\frac{1}{\rho}\nabla P + \nu \nabla^2 \mathbf{u}, \tag{B.31}$$

where $P = p - \frac{1}{2}\rho\Omega^2 r_\perp^2$ is the reduced pressure. The momentum equation is complemented with mass conservation: $\nabla \cdot \mathbf{u} = 0$.

Before proceeding it is convenient to introduce here the relevant dimensionless numbers which characterize the flow. These numbers indicate the relative importance of inertial, viscous and Coriolis forces in the momentum equation. For this purpose $\mathcal{L}$ and $\mathcal{U}$ represent the typical length and velocity scales of the flow, respectively. Together with the kinematic viscosity ν and system rotation Ω we can define three dimensionless numbers: the Reynolds number Re (ratio of inertial over viscous force), the *Rossby number Ro* (ratio of inertial over Coriolis force), and the *Ekman number Ek* (ratio of viscous over Coriolis force), defined as:

$$Re = \frac{\mathcal{U}\mathcal{L}}{\nu}, \qquad Ro = \frac{\mathcal{U}}{2\Omega\mathcal{L}}, \qquad Ek = \frac{\nu}{\Omega\mathcal{L}^2},$$

respectively. Obviously, only two of these dimensionless numbers are independent. Therefore, the dimensionless momentum equation contains two dimensionless numbers only: Ro and Ek (note that we have non-dimensionalized time by Ω^{-1} and the reduced pressure by $\rho\Omega\mathcal{U}\mathcal{L}$), and reads

$$\frac{\partial \mathbf{u}}{\partial t} + Ro\,(\mathbf{u} \cdot \nabla)\mathbf{u} + 2\mathbf{k} \times \mathbf{u} = -\nabla P + Ek\,\nabla^2\mathbf{u}, \tag{B.32}$$

where $\mathbf{k}$ is the unit vector in the direction of $\mathbf{\Omega}$.

Taylor–Proudman Theorem

When the Reynolds number is sufficiently large the flow becomes turbulent, and viscous effects are almost negligible in the bulk sufficiently far away from boundaries. In the boundary layers viscous effects are non-negligible and compete with the Coriolis force; see: 'Ekman boundary layers' for some further discussion. For large Reynolds number flow ($Re \gg 1$) in the bulk (far away from walls), combined with small Ekman number ($Ek \ll 1$), an almost (statistically) steady (mean) flow is fully characterized by the Rossby number. For rapid system rotation $Ro \ll 1$ (or $\mathcal{U} \ll \Omega\mathcal{L}$) we are allowed to ignore the advective and viscous contribution in (B.32). As we also assumed (statistical) steadiness of the (mean) flow the fluid motion is governed by

$$2\,\mathbf{k} \times \mathbf{u} = -\nabla P, \tag{B.33}$$

and fluid particle acceleration is fully determined by the balance between the pressure gradient and the Coriolis force. They both act perpendicular to the fluid motion, i.e., $\mathbf{u} \perp \nabla P$. This is known as the *geostrophic balance* and is of importance for atmospheric and large-scale oceanic (turbulent) flows.

An important relation for geostrophically-balanced flows can be obtained by taking the curl of the geostrophic balance (B.33). As $\nabla \times \nabla P = \mathbf{0}$ we obtain that $\nabla \times (\mathbf{k} \times \mathbf{u}) = \mathbf{0}$. Applying some vector algebra and taking into account flow incompressibility we arrive at the following relation:

$$\frac{\partial \mathbf{u}}{\partial z} = 0, \tag{B.34}$$

where we assign a coordinate system such that the rotation vector $\boldsymbol{\Omega}$ is aligned with the z-axis. From (B.34), known as the *Taylor–Proudman theorem* (Taylor 1917; Proudman 1916), we can immediately conclude that for rapidly rotating flows the velocity components will not change in the direction parallel with the rotation axis. This implies two-dimensional flow behavior in planes perpendicular to the rotation axis although the velocity component parallel to the rotation axis does not necessarily vanish (but takes on a constant value). Note again that we are considering (rapidly) rotating flows away from boundaries. Inside boundary layers the Taylor–Proudman theorem cannot be applied, see 'Ekman boundary layers'.

Taylor (1917) provided the experimental proof of this theorem by conducting laboratory experiments on steadily towing submerged obstacles through a rotating fluid. He observed that the flow was around the obstacle and not over it. In this way the obstacle carries a stagnant column of fluid as if the obstacle was extended over the full fluid layer. This stagnant volume of fluid is now known as the *Taylor column*. The emergence of Taylor columns are not restricted to these idealized experiments where the flow is more or less laminar. The formation of column-like structures are also observed in both laboratory experiments and in direct numerical simulations (DNS) of rotating turbulence. An example is shown in Fig. B.8 and concerns the decay of rotating turbulence behind a stroke of a vertically towed grid as visualized by the pearlescence technique in the experiments by Staplehurst et al. (2008) and reported by

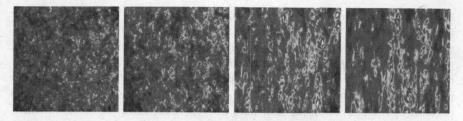

Fig. B.8 Experimental observation of columnar vortical structures in rotating turbulence experiments (Staplehurst et al. 2008). The pearlescence technique was used to visualize the vorticity field. Turbulence decays behind a stroke of a grid vertically towed through the fluid. From *left* to *right* snapshots are shown for increasing inertial times $\Omega/2\pi$ after the grid moved out of the field of view (instantaneous Rossby number is decreasing due to decreasing root mean square velocities promoting emergence of Taylor–Proudman like structures). From: Dalziel (2011)

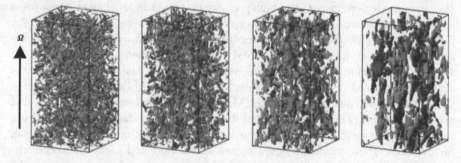

Fig. B.9 Isovorticity surfaces of rotating turbulence from direct numerical simulations. Snapshots from left to right are taken at increasing inertial times, and the formation of columnar vortical structures aligned with the background rotation is evident. From: Yoshimatsu et al. (2011)

Dalziel (2011). The formation of columnar vortical structures is evident. Yoshimatsu et al. (2011) reported recently the formation of vertically aligned vortical structures (parallel with the rotation axis) in a DNS. Isosurfaces of vorticity in a subregion of a 256^3 simulation grid points are shown in Fig. B.9.

Ekman Boundary Layers

In many geophysical and engineering applications it is reasonable to consider the bulk flow to be inviscid to good approximation. However, near domain boundaries (either a solid wall or an interface such as water-air) viscous effects likely become important and boundary layers emerge. For rotating flows these boundary layers have their specific characteristics and are not alike the classical Prandtl–Blasius boundary layers. The analysis of viscous boundary layers in rotating systems started at the end of the 19th century when it was observed by Nansen that the drift velocity of icebergs tend to be under a certain angle with respect to the wind (assumed to be constant on average for sufficient time). This observation challenged Ekman (1905) in the beginning of the last century to analyse the wind-driven motion and he succeeded to derive the boundary-layer structure at the ocean surface. Since then these are called

Ekman boundary layers. One should realize, however, that for turbulent flows (like in the atmosphere and oceans) one should not speak about viscous effects caused by molecular transport processes of momentum (like in laminar flows). The momentum transport is due to the turbulent eddies and the boundary-layer analysis is then actually based on a turbulent viscosity and mean boundary-layer profiles. In oceanography it is common to call the turbulent viscosity the 'Austausch' coefficient and its value depends on the position in the flow field as well as the direction of the turbulent momentum flux (perpendicular or parallel to the rotation axis). It includes therefore the statistical anisotropy of rotating turbulent flows.

For rapidly rotating (turbulent) flows the (mean) flow in the bulk is dictated by the geostrophic balance. As illustrated in the previous section the bulk might be populated with on average vertically-aligned cyclonic and anticyclonic vortical structures ('columns') with large-scale (mean) horizontal flows and associated pressure fields (low pressure regions in cyclones, high pressure regions in anticyclones). These pressure fields propagate more or less independently from the vertical coordinate to the boundary layers where friction becomes important. Friction will reduce the magnitude of the (mean) large-scale horizontal velocity components, and in its turn the Coriolis contribution. As the balance between Coriolis force and pressure gradient is now partly violated an inward swirl in cyclones and outward swirl in anticyclones is set up inside the Ekman boundary layer. Mass conservation then yields for horizontal convergence in cyclones a local upward flow from the boundary layer into the bulk and, the other way around, local downward flow from bulk into boundary layer in anticyclones. These mechanisms are known as *pumping* and *suction*, respectively, and is one of the distinctive differences of the Ekman boundary layer compared to the Prandtl–Blasius boundary layer: it actively influences the flow deep in the fluid bulk. Analysis of Ekman boundary layers is provided in textbooks (e.g., Kundu and Cohen 2004) and at higher technical level in the famous monograph by Greenspan (1968).

For practical purposes we align the system rotation with the (vertical) z-axis. A flat infinitely extended domain boundary (the bottom plate) is assumed to be in the (horizontal) xy-plane, see Fig. B.10. The (statistical) properties of the flow near this boundary are thus horizontally homogeneous and vertically inhomogeneous. The 'Austausch' coefficient generally has different values for the 'horizontal' and

Fig. B.10 A schematic representation of the Ekman boundary layer just above a horizontal plate. The angular velocity vector is pointing upwards (perpendicular to the bottom plate)

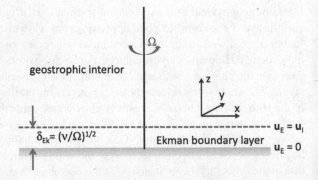

'vertical' diffusivity (A_h and A_v, respectively) implying

$$A\nabla^2 \mathbf{u} \rightarrow A_h \nabla_h^2 \mathbf{u} + A_v \frac{\partial^2 \mathbf{u}}{\partial z^2}, \tag{B.35}$$

where: $\nabla_h^2 = \frac{\partial^2}{\partial x^2} + \frac{\partial^2}{\partial y^2}$ is the Laplacian operator in a plane parallel to the bottom plate. For laminar flows $A_h = A_v = \nu$. Keeping the orientation of the domain boundary and the rotation axis in mind we can conclude that it is only needed to retain the z-derivative as changes in the velocity profile are largest in the direction perpendicular to the boundary. By again assuming (statistically) quasi-stationary flow with $Ro \ll 1$ and retaining the z-derivative of the viscous contribution we arrive at:

$$2\mathbf{k} \times \mathbf{u} = -\nabla P + Ek \frac{\partial^2 \mathbf{u}}{\partial z^2}, \tag{B.36}$$

with $Ek \ll 1$. In laboratory experiments and in ocean applications we find typically $Ek \approx 10^{-5}$. Equation (B.36) is a linear partial differential equation for the flow field $\mathbf{u}_E = (u_E, v_E, w_E)$ in the Ekman boundary layer and is the starting point for a mathematical analysis of this boundary layer, assuming an interior geostrophic flow field $\mathbf{u}_I = (u_I, v_I)$ and given boundary conditions. The general solution for the horizontal flow components is (assuming no-slip boundary conditions at the wall, see Fig. B.10):

$$u_E = u_I - [u_I \cos(z/\delta) + v_I \sin(z/\delta)] \exp(-z/\delta), \tag{B.37}$$

$$v_E = v_I + [u_I \sin(z/\delta) - v_I \cos(z/\delta)] \exp(-z/\delta). \tag{B.38}$$

Here, $\delta = \sqrt{Ek}$ is the dimensionless boundary layer thickness of the Ekman layer. The physical thickness is $\delta_{Ek} = \sqrt{\nu/\Omega}$ and is independent of the flow velocity. The Ekman boundary layer thickness is constant and thus independent of the particular flow configuration.

The Ekman boundary layers play an active role and they strongly influence the bulk properties of rotating turbulence, for example by pumping and suction of fluid into and out of the bulk of the fluid.

Inertial Waves

Rotating fluids support inertial waves, which are anisotropic and dispersive of nature. The propagation of internal waves is due to the restoring nature of the Coriolis force. It may be possible to describe rotating turbulence and its main features (for example, the formation of Taylor–Proudman-like coherent structures) in terms of interacting inertial waves and turbulence.

To quantify inertial wave motion we need to derive the relevant wave equation. In the limit $Ro \ll 1$ and $Ek \ll 1$ Eq. (B.31) reduces to

$$\frac{\partial \mathbf{u}}{\partial t} + 2\boldsymbol{\Omega} \times \mathbf{u} = -\frac{1}{\rho} \nabla P, \tag{B.39}$$

and taking the curl yields

$$\frac{\partial \omega}{\partial t} = 2(\mathbf{\Omega} \cdot \nabla)\mathbf{u}, \tag{B.40}$$

with $\omega = \nabla \times \mathbf{u}$. Differentiation of (B.40) with respect to time and subsequently taking its curl yields the wave equation:

$$\frac{\partial^2 (\nabla^2 \mathbf{u})}{\partial t^2} + 4(\mathbf{\Omega} \cdot \nabla)^2 \mathbf{u} = 0. \tag{B.41}$$

Solutions of this equation are plane waves of the form

$$\mathbf{u}(\mathbf{r}, t) = \mathcal{R}\left(\mathbf{A} \exp[i(\mathbf{l} \cdot \mathbf{r} - \tilde{\omega}t)]\right), \tag{B.42}$$

with $\mathcal{R}(\cdot\cdot)$ denoting the real part of a complex function, $\mathbf{A}$ the wave amplitude, i the imaginary unit, $\mathbf{l} = (l_x, l_y, l_z)$ the wave vector, and $\tilde{\omega}$ the angular frequency of Fourier mode $\mathbf{l}$. The latter is described by the dispersion relation $\tilde{\omega} = \pm 2\hat{\mathbf{l}} \cdot \mathbf{\Omega}$, with $\hat{\mathbf{l}}$ the unit vector in the direction of the wave vector $\mathbf{l}$ (with magnitude l). Phase and group velocities are

$$\mathbf{c}_p = \frac{\tilde{\omega}}{l}\hat{\mathbf{l}} = \pm\frac{2}{l}(\hat{\mathbf{l}} \cdot \mathbf{\Omega})\hat{\mathbf{l}}, \quad \text{and} \quad \mathbf{c}_g = \frac{\partial \tilde{\omega}}{\partial \mathbf{l}} = \pm\frac{2}{l}\left[\hat{\mathbf{l}} \times (\mathbf{\Omega} \times \hat{\mathbf{l}})\right], \tag{B.43}$$

respectively, with $\frac{\partial \tilde{\omega}}{\partial \mathbf{l}} = \left(\frac{\partial \tilde{\omega}}{\partial l_x}, \frac{\partial \tilde{\omega}}{\partial l_y}, \frac{\partial \tilde{\omega}}{\partial l_z}\right)$.

A few remarkable observations can be made. First of all we see that the absolute value of $\tilde{\omega} = \pm 2\hat{\mathbf{l}} \cdot \mathbf{\Omega}$ varies from zero to 2Ω and that low frequency waves have wavevectors almost perpendicular to $\mathbf{\Omega}$. Inertial waves are thus anisotropic. Moreover, the group and phase velocity of inertial waves are perpendicular: $\mathbf{c}_p \cdot \mathbf{c}_g = 0$. As a consequence, waves travelling in a certain direction $\hat{\mathbf{l}}$ (not parallel to $\mathbf{\Omega}$) propagate energy in a plane set up by $\hat{\mathbf{l}}$ and $\mathbf{\Omega}$ but perpendicular to $\hat{\mathbf{l}}$. More specifically, waves traveling in a direction perpendicular to the rotation axis propagate energy parallel to the rotation axis. Finally, low-frequency waves (with $\hat{\mathbf{l}}$ almost perpendicular to the rotation axis) propagate energy at a speed close to $2\Omega\mathcal{L}$ (with $\mathcal{L}$ either a typical length scale of turbulent patches or the size of a disturbance) in a direction parallel to the rotation axis. High-frequency waves (almost parallel with $\mathbf{\Omega}$) hardly propagate energy in that direction.

Conclusion

In this appendix we have introduced some of the basic mechanisms and phenomena needed for a phenomenological understanding of the behavior of rotating (turbulent) flows. For example, the formation of elongated structures in rotating turbulence, the global character of turbulence in the (Re, Ro)-plane, and the statistical anisotropy of many turbulence quantities when the flow is subjected to background rotation. Moreover, for confined rotating turbulence the Ekman boundary layers influence the turbulent flow in the bulk. Similar ingredients as discussed in this appendix may help elucidating the interpretation of more complicated aspects like, e.g., spectral scaling in rotating turbulence, structure function exponents and the behavior of velocity correlations in space and time. For a recent review on some of these topics, see Godeferd and Moisy (2015).

Problems

1. Provide an estimate of the Rossby number for large-scale atmospheric flows. Take, for example, a low-pressure system as a characteristic large-scale coherent structure.
2. Derive the components of the velocity field, see Eqs. (B.37) and (B.38), in the Ekman boundary layer for the geometry discussed in the 'Ekman boundary layer' section above.
3. Consider rotating turbulence and as observer you move with the co-rotating frame of reference. Give the equations for the kinetic energy of the mean flow and the kinetic energy of the turbulence. Explain the presence/absence of contributions directly related with system rotation (thus terms containing Ω_i).
4. Consider now the Boussinesq equation in a rotating frame of reference. Gravity and the background rotation are parallel. Derive the analogon of the Taylor–Proudman theorem for the rotating Boussinesq equation. Give a physical explanation. This relation is known as the 'thermal wind balance' in the geophysical fluid dynamics community.

B.5 Drag Reduction by Polymer Additives

Small amounts of certain polymer additives to fluids can achieve a significant reduction of friction drag, which is know as the *Toms effect* (Toms 1948; Virk 1975). This is quite a remarkable effect, as the addition of a polymer to the solvent somewhat *increases* the fluid viscosity, while a significant *reduction* of the friction occurs, even up to 70 % with respect to that of the original solvent. Due to the viscoelastic properties of the polymers the fluid can no longer be considered as Newtonian. The effect occurs for various polymers; polyethylene-oxide (PEO) and polyacrylamide (PAM) are commonly used water-soluble polymers, while polyisobutylene is a commonly used drag reducing agent in hydrocarbon fluids (viz., oil). Polymer drag reduction has found several practical applications in reducing the economic costs of fluid transport through pipes; perhaps the most well-known application is that of the increase of oil transport through the trans Alaska pipe line (Burger et al. 1982). However, special care must be taken to avoid degradation of the polymers, for instance as the result of intense shearing (which typically occurs in certain pumps), by which the drag reducing effect is lost.

This section provides a brief description and analysis of this effect, and for further reading we refer to reviews by Lumley (1969), Nieuwstadt and den Toonder (2001), and White and Mungal (2008).

Early studies of polymer drag reduction involve observations of the pressure drop as a function of the flow rate in a pipe flow. For a pipe flow, the friction factor

Fig. B.11 The friction factor λ for pipe flow, plotted as $\lambda^{-1/2}$ versus $\mathrm{Re}\sqrt{\lambda}$, showing the onset and trajectories of polymer drag reduction for different polymers (PEO: polyethylene-oxide; PAM: polyacrylamide, where M indicates the polymer molecular weight in g/mol) at various concentrations c (in *weight parts per million*, or wppm) and pipe diameters D (in mm). The *lines represent* the laminar friction law ($\lambda = 64/\mathrm{Re}$), the Prandtl–Kármán law (6.35) with $A = 1.884$ and $B = 0.331$, and the drag reduction asymptote (B.44). After: Virk (1975); data represented by the symbols * and ⋆ are from Ptasinski et al. (2001)

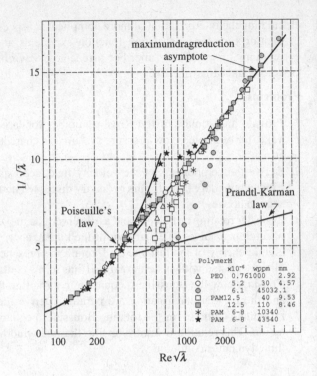

λ, that defines a relation between the bulk flow rate and pressure drop, is given in (6.34). Figure B.11 shows a Prandtl–Kármán plot of $\lambda^{-1/2}$ versus $\mathrm{Re}\sqrt{\lambda}$, where the Reynolds number is based on the flow rate Q $(=\frac{\pi}{4}D^2 U_b)$ and the viscosity of the solvent.

At low Reynolds numbers ($\mathrm{Re}\sqrt{\lambda} < 400$) all flows are in the laminar flow regime, i.e., $\lambda = 64/\mathrm{Re}$. When the flow rate increases, a transition to turbulence occurs, which is especially clear for the solid dots in Fig. B.11. Initially, the data follow the Prandtl–Kármán law (6.35) that is valid for a Newtonian fluid, until a certain flow rate, where the data diverge to lower values of the friction factor; this is the onset of polymer drag reduction. Experiments indicate that the onset of drag reduction occurs only after a certain wall shear stress has been exceeded, which means that the onset of drag reduction depends on the pipe diameter. Hence for pipes with a small diameter the onset of drag reduction occurs for lower Reynolds number, and can even occur directly at the transition to turbulence, as can be seen in Fig. B.11. After onset, the drag reduction increases with increasing Reynolds numbers, where it follows approximately a straight line in the Prandtl–Kármán plot, until it reaches another straight line that is described by[2]:

---

[2]Here we express this asymptote for the Darcy friction factor λ, whereas the asymptote by Virk (1975) was given for the Fanning friction factor that is equal to $\lambda/4$.

$$\frac{1}{\sqrt{\lambda}} = 9.5 \log \left(\text{Re} \sqrt{\lambda} \right) - 16.2. \tag{B.44}$$

This is an empirical asymptote for which a further increase in the concentration of the polymer does not lead to a further reduction of the drag (Virk 1975). Equation (B.44) is commonly referred to as the 'maximum drag reduction asymptote' or *Virk asymptote*.

The common thought is that the onset of drag reduction occurs when the polymer time scale T_P, i.e., the average time it takes for a stretched polymer to return to its coiled configuration, becomes comparable to the near-wall turbulence time scale ν/u_*^2. The ratio of these two time scales is defined as the wall-shear *Weissenberg number*:

$$\text{Wi}_* = \frac{T_P u_*^2}{\nu}.$$

However, it should be noted that the onset of drag reduction is also clearly dependent on the polymer concentration (see Fig. B.11), which is not reflected in Wi_*.

In Sect. 6.3 the friction law (6.35) for Newtonian fluids was derived from the logarithmic law (6.15) by assuming that it could be used as an approximation for the velocity profile across the entire pipe cross section. The reverse procedure was used by Virk et al. (1967) to derive the following ultimate profile from the maximum drag reduction asymptote (B.44):

$$u^+ = 11.7 \ln \left(y^+ \right) - 17.0. \tag{B.45}$$

Figure B.12 shows experimental data of the measured velocity profiles for various flows with polymer additives. In the viscous sublayer, the velocity profile does not show any changes from (6.21) for Newtonian fluids. Further from the wall a logarithmic region is observed in case of the drag-reducing flows, with the same slope as in (6.21) for a Newtonian fluid. In between the viscous region and the logarithmic region, the profiles for the drag-reducing flows seem to follow the ultimate profile. The result appears as a thickened buffer layer in which the velocity increases by an amount $\Delta\Pi/k$, which is called the *effective slip* relative to the solvent velocity.

This is also evident from measurements of the turbulence intensities, as shown in Fig. B.13. The addition of the polymers causes the peak in the axial velocity fluctuations to broaden and move outward from $y^+ = 20$ to $y^+ = 50$, while the peak value is first increased, but returns to an amplitude comparable to that for a Newtonian fluid when the polymer concentration reaches the value for maximum drag reduction (at 435 wppm in Fig. B.13). A similar outward movement can be observed for the radial velocity fluctuations, while the amplitude of the velocity fluctuations continuously reduces with increasing polymer concentration.

Hence, the drag reduction phenomenon may be attributed to changes in the buffer layer. With increasing drag reduction, the effective slip $\Delta\Pi/k$ increases, until the buffer layer, with the ultimate profile (B.45), would span the entire pipe cross section

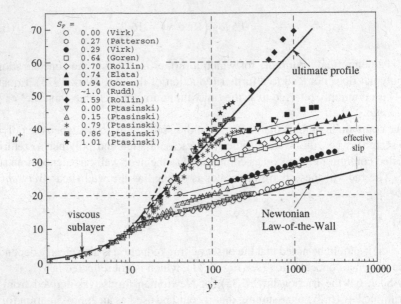

Fig. B.12 The mean velocity profiles for various cases of polymer drag reduction. The *solid lines* represent the viscous sublayer and logarithmic profile in (6.21), and the ultimate profile in (B.45). The profiles for drag-reducing flows have logarithmic profiles with the same slope as for a Newtonian fluid, but with an offset, or *effective slip*. Experimental details and references can be found in the paper by Virk (1975). The parameter S_F equals the fractional increase in flow rate Q' relative to the Newtonian flow rate Q at equal pressure drop, i.e., $S_F = \sqrt{\lambda/\lambda'} - 1 = Q'/Q - 1$. From: Virk (1975), with data of Ptasinski et al. (2001) included

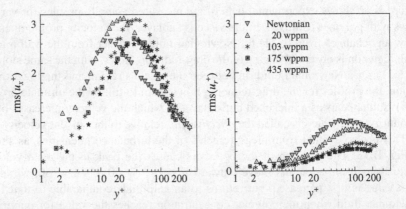

Fig. B.13 The root mean square of the axial (*left*) and radial (*right*) components of the fluctuating velocity (normalized by the wall friction velocity) in a pipe flow for various polymer concentrations in weight parts per million (wppm). Data were obtained through LDV measurements in a pipe flow at Re $= 10^4$. Symbols correspond to those in Figs. B.11 and B.12. From: Ptasinski et al. (2001)

(with the exception of the viscous sublayer near the wall), as illustrated in Fig. B.12. The enlarged buffer layer in drag-reducing flow is referred to as the *elastic sublayer*.

It was noted by Lumley (1969) that the profiles for the drag-reducing flows retain essentially the basic structure as for the flow of a Newtonian fluid, which were described in Sect. 6.3. Hence, if one assumes that the velocity profile in the viscous sublayer (6.18) matches the logarithmic profile (6.15) at $y^+ = y_v^+$, then one can write

$$\frac{\Pi}{k} = y_v^+ - \frac{1}{k} \ln \left(y_v^+ \right). \tag{B.46}$$

Then, following the scaling arguments mentioned in Sect. 6.3 we can formulate a *friction law*:

$$\frac{u_0}{u_*} = \frac{1}{k} \ln \left(\frac{R u_*}{\nu} \right) + y_v^+ - \frac{1}{k} \ln \left(y_v^+ \right) - C, \tag{B.47}$$

where $R(= \frac{1}{2} D)$ is the pipe radius and C a constant. From this expression it follows that if u_* is decreased at equal u_0 (that is, the friction coefficient is *reduced* as a result of drag reduction), then y_v^+ is *increased*. This then implies that a drag reduction indeed must appear as a thickened sublayer.

For a further analysis of the physical aspects of drag reducing flows we return to the basic equations of motions for turbulent flow. To account for the non-Newtonian behavior of the polymers, we modify the stress tensor σ_{ij}, defined in (2.8), by adding a term that represents the *polymeric stress* τ_{ij}^P:

$$\sigma_{ij} = -p\delta_{ij} + \mu \left(\frac{\partial u_i}{\partial x_j} + \frac{\partial u_j}{\partial x_i} \right) + \tau_{ij}^P. \tag{B.48}$$

Substitution of (B.48) in (2.7) yields a modified momentum equation (2.10). Then separating all turbulent quantities in mean and fluctuating parts, as explained in Sect. 5.3, yields the following Reynolds-averaged Navier-Stokes equations:

$$\frac{D\overline{u_i}}{Dt} = -\frac{1}{\rho_0} \frac{\partial \overline{p}}{\partial x_i} + \frac{\partial}{\partial x_j} \left(\nu \frac{\partial \overline{u_i}}{\partial x_j} - \overline{u_i' u_j'} + \frac{1}{\rho_0} \overline{\tau_{ij}^P} \right), \tag{B.49}$$

cf. (5.20). In order to solve this equation, we need a closure relation for $\overline{\tau_{ij}^P}$, in addition to a closure relation for $\overline{u_i' u_j'}$.

For fully developed pipe flow, the left-hand side of (B.49) vanishes and the total stress τ_t increases linearly with the radial distance r from the pipe centerline. Following the notation of (6.9), with: $y = R - r$, where R is the pipe radius, we find:

$$\frac{\tau_t}{\rho_0} = -\overline{u'v'} + \nu \frac{\partial \overline{u}}{\partial y} - \frac{\overline{\tau_{rx}^P}}{\rho_0} = u_*^2 \left(1 - \frac{y}{R} \right), \tag{B.50}$$

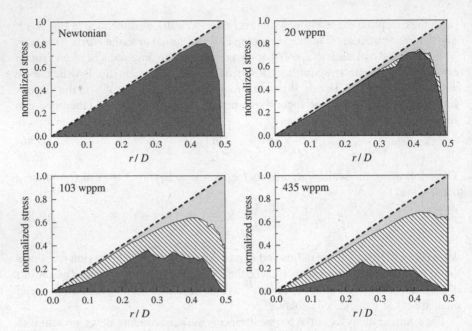

Fig. B.14 Radial profiles of the total shear stress normalized by the wall friction velocity in a turbulent pipe flow at Re = 10^4 for various polymer concentrations in weight parts per million (wppm). The *light gray* region corresponds to the viscous stress, while the *dark gray* region corresponds to the Reynolds stress; the difference is the polymer stress, represented by the hashed region. The graph for the Newtonian fluid case (*top left*) is equivalent to Fig. 6.4. Data of Ptasinski et al. (2001), corresponding to those in Fig. B.13

where $\overline{u}$ is the mean axial velocity. The Reynolds stress $\overline{u'v'}$ and viscous stress $\nu \partial \overline{u}/\partial y$ can be measured directly, e.g., by means of two-component LDV. Then, the polymeric stress can be inferred from the deficit between the total stress and the measured Reynolds stress and viscous stress. Figure B.14 shows the contributions of these three terms to the total stress; the polymer stress is inferred from the deficit between the total stress and the measured viscous stress and Reynolds stress.

Following the steps explained in Sect. 7.1, we obtain the following expression for the kinetic energy of the mean flow in a pipe:

$$0 = \underbrace{-\frac{1}{\rho_0}\overline{u}\frac{d\overline{p}}{dx}}_{P_u} + \underbrace{\frac{1}{r}\frac{\partial}{\partial r}\left[r\overline{u}\left(-\overline{u'_x u'_r} + \nu\frac{d\overline{u}}{dr} + \frac{1}{\rho_0}\overline{\tau^P_{rx}}\right)\right]}_{T_u} + \underbrace{\overline{u'_x u'_r}\frac{d\overline{u}}{dr}}_{D_u} - \underbrace{\nu\left(\frac{d\overline{u}}{dr}\right)^2}_{E_s} - \underbrace{\frac{1}{\rho_0}\overline{\tau^P_{rx}}\frac{d\overline{u}}{dr}}_{E_p},$$

(B.51)

where P_u is the pressure work, T_u transport of mean-flow energy by: (i) Reynolds stress, (ii) viscous stress, and (iii) mean polymeric stress, D_u the deformation work by Reynolds stress, E_s viscous dissipation, and E_p the polymeric dissipation. As previously explained in Sect. 7.1, the terms in T_u are divergences of the energy flux, and only redistribute the energy, while D_u is always negative and appears as a source term in the equation for the turbulent kinetic energy. The term E_s is typically very small, and is usually neglected. The new term E_p is also a loss term that is considerable (as can be inferred from Fig. B.14). This implies that most of the energy is transferred directly to the polymers, and not through the route of a turbulence energy cascade.

A solid theoretical basis that explains and can predict the drag reduction for polymer additives is still lacking. Several theories have been proposed, but none of them is without criticism. It is however instructive to review the mechanism described by Lumley (1969).

According to Lumley (1969) the drag reduction effect can be explained in terms of the scaling of the various wall turbulence regions defined in Sect. 6.2. The basic assumption is that the polymers are expanded in the flow outside the viscous sub-layer due to the fluctuating rate-of-strain, which causes an increase of the effective viscosity. This changes the scaling behavior of the wall turbulence, as illustrated in Fig. B.15. In this figure are plotted the length scale $\mathcal{L}^+$ of the energy-containing eddies (6.13) and the length scale ℓ_ϵ^+ of the dissipative eddies, both in dimensionless wall units (6.19). The peak of the turbulent kinetic energy spectrum may be expected to occur at $\kappa y = 1$, so that $\mathcal{L}^+ = y^+$, while the peak in the dissipation spectrum occurs at $\kappa \eta = 0.2$ (see Problem 14 in Sect. 9.6). Since production and dissipation are mainly balanced in the wall region (see Sect. 7.2), it follows that

$$\epsilon = \frac{u_*^3}{ky}, \quad \text{and thus:} \quad \ell_\epsilon^+ = 5\left(ky^+\right)^{1/4}, \quad \text{with:} \quad 5\eta^+ = 5\frac{u_*}{\nu}\left(\frac{\nu^3}{\epsilon}\right)^{1/4}. \quad (B.52)$$

Turbulence can only maintain itself when $\mathcal{L} \gg \ell_\epsilon$, which is indicated by the hashed region in Fig. B.15. This then occurs for $y^+ > 6.3$, which is consistent with the thickness $\delta_v^+ = 5$ of the viscous sublayer (see Sect. 6.2).

Since the polymer molecules do not become stretched in the viscous sublayer, i.e., the viscosity in the viscous sublayer is effectively unchanged, the normalization

Fig. B.15 Scaling relation in the viscous and inertial sublayers, with and without polymers according to Lumley (1969)

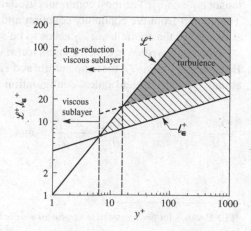

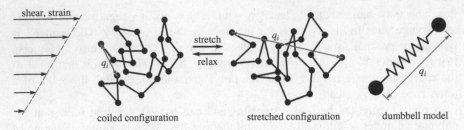

shear, strain

q_i

stretch

relax

q_i

q_i

coiled configuration   stretched configuration   dumbbell model

Fig. B.16 Schematic of polymer stretch and relaxation in shear flow. The polymer extension is characterized by the change in q_i. The polymer stretch can be modeled by a single dumbbell model. After: White and Mungal (2008)

in terms of wall units in Fig. B.15 remains the same. Outside the viscous sublayer, the viscosity is increased. The dissipation is determined by the energy-containing eddies (and not by the dissipative eddies), so that the local value of the viscosity simply shifts upward the line for ℓ_ϵ (reducing the turbulent flow region to the gray region), so that the intersection occurs at a larger distance from the wall. In the case of maximum drag reduction the sublayer thickness would have increased to reach the pipe center.

The concept of Lumley's theory (1969) is attractively simple, but remains qualitative. Also note that it predicts a thickening of the *viscous sublayer*, while we have seen earlier from experimental data that it is the *buffer layer* that becomes thickened (e.g., see Fig. B.12).

Modern investigations of polymer drag reduction utilize numerical simulations of the flow. Constitutive relations are derived from modeling the polymers as a single *dumbbell*, i.e., two beads connected by an elastic spring; see Fig. B.16. The dynamics of the polymers are then represented by the evolution of the end-to-end vector q_i connecting the beads. The beads are interacting through hydrodynamic forces with the fluid motion, while a diffusive process accounts for Brownian fluctuations in the configuration of the polymers, represented as the phase-averaged configuration tensor $c_{ij} = \overline{q_i q_j}$. The most commonly used dumbbell model is the FENE-P model.[3] Then, apart from the continuity equation and momentum equations, and additional equation for the evolution of c_{ij} needs to be solved.

When q_i is written in dimensionless form, i.e., $\tilde{q}_i = q_i/\sqrt{k_B T/H}$, where k_B is the Boltzmann constant, T the temperature, and H the spring constant, then the evolution equation for the dimensionless configuration tensor $\tilde{c}_{ij}$ and polymer stress read:

$$\frac{D\tilde{c}_{ij}}{Dt} = \tilde{c}_{kj}\frac{\partial u_k}{\partial x_i} + \tilde{c}_{ik}\frac{\partial u_k}{\partial x_j} - \frac{1}{T_P}\frac{\tau_{ij}^P}{\Sigma_0}, \quad \text{and:} \quad \tau_{ij}^P = \Sigma_0\left(-\delta_{ij} + \frac{\tilde{c}_{ij}}{1-\frac{\tilde{c}_{kk}}{\tilde{q}_0^2}}\right), \quad (B.53)$$

---

[3]FENE stands for *finite extensible non-linear elastic* and P for the closure proposed by Peterlin.

with: $\Sigma_0 = nk_BT$. Here $T_P = \zeta/4H$ is the relaxation time of the dumbbell, where ζ is the Stokes friction coefficient of the beads. Further details can be found in the literature (e.g., Sureshkumar et al. 1997; Ptasinski et al., 2003; White and Mungal 2008).

The numerical simulations give results that resemble those of the experimental observations presented in the figures above, at least for low polymer concentrations. As with any other closure models reviewed previously in this book, there are several assumptions that underlie the closure hypothesis, so it can be expected that this model is imperfect. For example, the single dumbbell model is rather simplistic for the representation of very long polymer chains, and polymer-polymer interactions are not included. Also, the dumbbell model assumes that the drag reduction is related to the elasticity of the polymers, while recent measurements and numerical simulations have shown that stiff polymers, or fibers, can also achieve substantial drag reduction (Gillissen et al. 2008).

References

Acheson, D. J. 1990 *Elementary Fluid Mechanics*. Oxford, UK: Clarendon.

Adrian, R. J. 1979 Conditional eddies in isotropic turbulence. *Phys. Fluids* **22**, 2065–2070.

Adrian, R. J. 2007 Hairpin vortex organization in wall turbulence. *Phys. Fluids* **19**, 041301.

Adrian, R. J., Meinhart, C. D. & Tomkins, C. D. 2000 Vortex organization in the outer region of the turbulent boundary layer. *J. Fluid Mech.* **422**, 1–54.

Adrian, R. J. & Westerweel, J. 2011 *Particle Image Velocimetry*. Cambridge University Press.

Atsavapranee, P. & Gharib, M. 1997 Structures in stratified plane mixing layers and the effects of cross-shear. *J. Fluid Mech.* **342**, 53–86.

Auton, T. R. 1987 The lift force on a spherical body in a rotational flow. *J. Fluid Mech.* **183**, 199–218.

Auton, T. R., Hunt, J. C. R. & Prudhomme, M. 1988 The force exerted on a body in inviscid unsteady non-uniform rotational flow. *J. Fluid Mech.* **197**, 241.

Avila, K., Moxey, D., de Lozar, A., Avila, M., Barkley, D. & Hof, B. 2011 The onset of turbulence in pipe flow. *Science* **333**, 192–196.

Batchelor, G. K. 1953 *The Theory of Homogeneous Turbulence*. Cambridge: Cambridge University Press.

Batchelor, G. K. 1967 *An Introduction to Fluid Mechanics*. Cambridge University Press.

Bechert, B. W., Bruse, M., Hage, W. & Meyer, R. 2000 Fluid mechanics of biological surfaces and their technological application. *Naturwissenschaften* **87**, 157–171.

Bergé, P., Pomeau, Y. & Vidal, C. 1984 *Order within chaos – Towards a deterministic approach to turbulence*. New York, NY: John Wiley & Sons.

Boersma, B. J., Brethouwer, G. & Nieuwstadt, F. T. M. 1998 A numerical investigation on the effect of the inflow conditions on the self-similar region of a round jet. *Phys. Fluids* **10**, 899–909.

Britter, R. E., Hunt, J. C. R. & Mumford, J. C. 1979 The distortion of turbulence by a circular cylinder. *J. Fluid Mech.* **92**, 269–301.

Brown, G. L. & Roshko, A. 1974 On density effects and large structure in turbulent mixing layers. *J. Fluid Mech.* **64**, 775–816.

Burger, E. D., Munk, W. R. & Wahl, H. A. 1982 Flow increase in the trans alaska pipeline through use of a polymeric drag-reducing additive. *J. Petrol. Technol.* **34**, 377–386.

Cabot, W. & Moin, P. 1999 Approximate wall boundary conditions in the large-eddy simulation of high-Reynolds number flow. *Flow Turbul. Combust.* **63**, 269–291.

Cantwell, B., Coles, D. & Dimotakis, P. 1978 Structure and entrainment in the plane of symmetry of a turbulent spot. *J. Fluid Mech.* **87**, 641–672.

Caporaloni, M., Tampieri, F., Trombetti, F. & Vittori, O. 1975 Transfer of particles in nonisotropic air turbulence. *J. Atmos. Sci.* **32**, 565–568.

© Springer International Publishing Switzerland 2016

F.T.M. Nieuwstadt et al., *Turbulence*, DOI 10.1007/978-3-319-31599-7

Catrakis, H. J. & Dimotakis, P. E. 1996 Scale distributions and fractal dimensions in turbulence. *Phys. Rev. Lett.* **77**, 3795–3798.

Champagne, F. H., Harris, V. G. & Corrsin, S. 1970 Experiments on nearly homogeneous turbulent shear flow. *J. Fluid Mech.* **41**, 81–139.

Chen, J., Fan, A., Zou, J., Engel, J. & Liu, C. 2003 Two-dimensional micromachined flow sensor array for fluid mechanics studies. *J. Aerospace Eng.* **16**, 85–97.

Chong, M. S., Perry, A. E. & Cantwell, B. J. 1990 A general classification of three-dimensional flow fields. *Phys. Fluids A* **2**, 765–777.

Comte-Bellot, G. 1976 Hot-wire anemometry. *Annu. Rev. Fluid Mech.* **8**, 209–231.

Comte-Bellot, G. & Corrsin, S. 1966 The use of a contraction to improve the isotropy of grid-generated turbulence. *J. Fluid Mech.* **25**, 657–682.

Comte-Bellot, G. & Corrsin, S. 1971 Simple Eulerian time correlation of full- and narrow-band velocity signals in grid-generated, 'isotropic' turbulence. *J. Fluid Mech.* **48**, 273–337.

Csanady, G. T. 1973 *Turbulent Diffusion in the Environment*. Dordrecht, NL: Kluwer Academic Publishers.

Dalziel, S. B. 2011 The twists and turns of rotating turbulence. *J. Fluid Mech.* **666**, 1–4.

Davidson, P. A. 2004 *Turbulence – An Introduction for Scientists and Engineers*. Oxford, UK: Oxford University Press.

Davidson, P. A. 2013 *Turbulence in Rotating, Stratified and Electrically Conducting Fluids*. Cambridge, UK: Cambridge University Press.

del Álamo, J. C., Jiménez, J., Zandonade, P. & Moser, R. D. 2004 Scaling of the energy spectra of turbulent channels. *J. Fluid Mech.* **500**, 135–144.

Drazin, P. G. & Reid, W. H. 1981 *Hydrodynamic Stability*. Cambridge University Press.

Eggels, J. G. M., Unger, F., Weiss, M. H., Westerweel, J., Adrian, R. J., Friedrich, R. & Nieuwstadt, F. T. M. 1994 Fully developed turbulent pipe flow: a comparison between direct numerical simulation and experiment. *J. Fluid Mech.* **268**, 175–209.

Ekman, V. W. 1905 On the influence of the Earth's rotation on ocean currents. *Ark. Mat. Astron. Fys.* **2**, 1–53.

Ewing, D., Frohnapfel, B., George, W. K., Pedersen, J. M. & Westerweel, J. 2007 Two-point similarity in the round jet. *J. Fluid Mech.* **577**, 309–330.

Faber, T. E. 1995 *Fluid Dynamics for Physicists*. Cambridge, UK: Cambridge University Press.

Faisst, H. & Eckhardt, B. 2003 Traveling waves in pipe flow. *Phys. Rev. Lett.* **91**, 224502.

Faisst, H. & Eckhardt, B. 2004 Sensitive dependence on initial conditions in transition to turbulence in pipe flow. *J. Fluid Mech.* **504**, 343–352.

Fleagle, R. G. & Businger, J. A. 1980 *An Introduction to Atmospheric Physics*, 2nd edn., *International Geophysics Series*, vol. 25. New York, NY: Academic Press.

Frisch, U. 1995 *Turbulence: The Legacy of A. N. Kolmogorov*. Cambridge University Press.

Fukushima, C., Aanen, L. & Westerweel, J. 2002 Investigation of the mixing process in an axisymmetric turbulent jet using PIV and LIF. In *Laser Techniques for Fluid Mechanics* (ed. R. J. Adrian *et al.*), pp. 339–356. Berlin: Springer.

Gavarini, M. I., Bottaro, A. & Nieuwstadt, F. T. M. 2004 The initial stage of transition in pipe flow: role of optimal base-flow distortions. *J. Fluid Mech.* **517**, 131–165.

George, W. K. 1989 The self-preservation of turbulent flows and its relation to initial conditions and coherent structures. In *Advances in Turbulence* (ed. W. K George & R. Arndt), pp. 39–73. New York, NY: Hemisphere.

Gibson, M. M. 1963 Spectra of turbulence in a round jet. *J. Fluid Mech.* **15**, 161–173.

Gillissen, J. J. J., Boersma, B. J., Mortensen, N. A. & Andersson, H. I. 2008 Fibre-induced drag reduction. *J. Fluid Mech.* **602**, 209–218.

Godeferd, F. S. & Moisy, F. 2015 Structure and dynamics of rotating turbulence: a review of recent experimental and numerical results. *Appl. Mech. Rev.* **67**, 030802.

Goldstein, R. J. 1996 *Fluid Mechanics Measurement*, 2nd edn. Washington DC: Taylor & Francis.

Gollub, J. P. & Benson, S. V. 1980 Many routes to turbulent convection. *J. Fluid Mech.* **100**, 449–470.

Grant, H. L., Stewart, R. W. & Moilliet, A. 1962 Turbulence spectra from a tidal channel. *J. Fluid Mech.* **12**, 241–268.

Greenspan, H. P. 1968 *The Theory of Rotating Fluids*. Cambridge, UK: Cambridge University Press.

Gronald, G. & Derksen, J. J. 2011 Simulating turbulent swirling flow in a gas cyclone: A comparison of various modeling approaches. *Powder Technol.* **205**, 160–171.

Head, M. R. & Bandyopadhyay, P. R. 1981 New aspects of turbulent boundary-layer structure. *J. Fluid Mech.* **107**, 297–338.

Hinze, J. O. 1975 *Turbulence*, 2nd edn. New York: McGraw-Hill.

Hof, B., de Lozar, A., Kuik, D. J. & Westerweel, J. 2008 Repeller or attractor? selecting the dynamical model for the onset of turbulence in pipe flow. *Phys. Rev. Lett.* **101**, 214501.

Hof, B., van Doorne, C. W. H., Westerweel, J., Nieuwstadt, F. T. M., Faisst, H., Eckhardt, B., Wedin, H., Kerswell, R. R. & Waleffe, F. 2004 Experimental observation of nonlinear traveling waves in turbulent pipe flow. *Science* **305**, 1594–1598.

Hof, B., Westerweel, J., Schneider, T. M. & Eckhardt, B. 2006 Finite lifetime of turbulence in shear flows. *Nature* **443**, 59–62.

Hopfinger, E. J., Browand, F. K. & Gagne, Y. 1982 Turbulence and waves in a rotating tank. *J. Fluid Mech.* **125**, 505–534.

Howe, M. S. 2003 *Theory of Vortex Sound*. Cambridge University Press.

Hunt, J. C. R., Eames, I. & Westerweel, J. 2006 Mechanics of inhomogeneous turbulence and interfacial layers. *J. Fluid Mech.* **554**, 499–519.

Hunt, J. C. R., Wray, A. A. & Moin, P. 1988 Eddies, streams, and convergence zones in turbulent flows. In *Proc. CTR Summer Program*, pp. 193–208. CTR.

Hussein, H. J., Capp, S. P. & George, W. K. 1994 Velocity measurements in a high-Reynolds-number, momentum-conserving, axisymmetric, turbulent jet. *J. Fluid Mech.* **258**, 31–75.

Ishihara, T., Gotoh, T. & Kaneda, Y. 2009 Study of high-Reynolds number isotropic turbulence by direct numerical simulation. *Annu. Rev. Fluid Mech.* **41**, 165–180.

Joseph, D. D. & Ocando, D. 2002 Slip velocity and lift. *J. Fluid Mech.* **454**, 263–286.

Kim, J. & Antonia, R. A. 1993 Isotropy of the small scales of turbulence at low Reynolds number. *J. Fluid Mech.* **251**, 219–238.

Kim, J., Moin, P. & Moser, R. 1987 Turbulence statistics in fully developed channel flow at low reynolds number. *J. Fluid Mech.* **177**, 133–166.

Kistler, A. L. & Vrebalovich, T. 1966 Grid turbulence at large Reynolds numbers. *J. Fluid Mech.* **26**, 37–47.

Kline, S. J., Reynolds, W. C., Schraub, F. A. & Runstadler, P. W. 1967 The structure of turbulent boundary layers. *J. Fluid Mech.* **30**, 741–773.

Kolmogorov, A. N. 1962 A refinement of previous hypotheses concerning the local structure of turbulence in a viscous incompressible fluid at high reynolds number. *J. Fluid Mech.* **13**, 82–85.

Kolmogorov, A. N. 1991 The local structure of turbulence in incompressible viscous fluid for very large Reynolds numbers. *Proc. R. Soc. Lond. A* **434**, 9–13.

Kuik, D. J., Poelma, C. & Westerweel, J. 2010 Quantitative measurement of the lifetime of localized turbulence in pipe flow. *J. Fluid Mech.* **645**, 529–539.

Kundu, P. K. & Cohen, I. M. 2004 *Fluid Mechanics*, 3rd edn. Amsterdam, NL: Elsevier Academic.

Landahl, M. T. & Mollo-Christensen, E. 1986 *Turbulence and Random Processes in Fluid Mechanics*. Cambridge, UK: Cambridge University Press.

Landau, L. D. & Lifshitz, E. M. 1959 *Fluid Mechanics, A Course of Theoretical Physics*, vol. 6. Pergamon Press.

Langelandsvik, L. I., Kunkel, G. J. & Smits, A. J. 2008 Flow in a commercial steel pipe. *J. Fluid Mech.* **595**, 323–339.

Laufer, J. 1954 The structure of turbulence in fully developed pipe flow. Report 1174. NACA.

Lesieur, M. 2008 *Turbulence in Fluids*, 4th edn. Springer-Verlag.

Lighthill, M. J. 1952 On the sound generate aerodynamically, I general theory. *Proc. R. Soc. Lond. A* **211**, 564–587.

Lorenz, E. N. 1963 Deterministic nonperiodic flow. *J. Atmos. Sci.* **20**, 130–141.

Ludwieg, H. & Tillmann, W. 1950 Investigations of the wall-shearing stress in turbulent boundary layers. TM 1285. NACA, Washington, DC.

Lumley, J. L. 1969 Drag reduction by additives. *Annu. Rev. Fluid Mech.* **1**, 367–384.

Mandelbrot, B. B. 1977 *Fractals: Form, Change and Dimension*. San Francisco, CA: Freeman.

Mansour, N. N., Kim, J. & Moin, P. 1988 Reynolds-stress and dissipation-rate budgets in a turbulent channel flows. *J. Fluid Mech.* **194**, 15–44.

Maxey, M. R. & Riley, J. J. 1983 Equation of motion for a small rigid sphere in a nonuniform flow. *Phys. Fluids* **26**, 883.

McComb, W. D. 1990 *The Physics of Fluid Turbulence*. Oxford, UK: Oxford University Press.

McKeon, B. J., Li, J., Jiang, W., Morrison, J. F. & Smits, A. J. 2004 Further observations on the mean velocity distribution in fully developed pipe flow. *J. Fluid Mech.* **501**, 135–147.

McLaughlin, J. B. 1993 The lift on a small sphere in wall-bounded linear shear flows. *J. Fluid Mech.* **246**, 249–265.

Mei, R. 1996 Velocity fidelity of flow tracer particles. *Exp. Fluids* **22**, 1–13.

Merzkirch, W. 1987 *Flow Visualization*, 2nd edn. Orlando: Academic.

Moin, P. & Kim, J. 1982 Numerical investigation of turbulent channel flow. *J. Fluid Mech.* **118**, 341–377.

Moin, P. & Mahesh, K. 1998 Direct numerical simulation: a tool in turbulence research. *Annu. Rev. Fluid Mech.* **30**, 539–578.

Monin, A. S. & Yaglom, A. M. 1973 *Statistical Fluid Mechanics: Mechanics of Turbulence (Vol. I)*. Cambridge, MA: MIT Press.

Moore, P. D. 2009 Aeroacoustics of compressible subsonic jets: Direct Numerical Simulation of a low Reynolds number subsonic jet and the associated sound field. Ph.D. thesis, Delft University of Technology.

Moxey, D. & Barkley, D. 2010 Distinct large-scale turbulent-laminar states in transitional pipe flow. *P. Natl. Acad. Sci.* **107**, 8091–8096.

Mungal, M. G. & Hollingsworth, D. K. 1989 Organized motion in a very high Reynolds number jet. *Phys. Fluids A* **1**, 1615–1623.

Nieuwstadt, F. T. M. & den Toonder, J. M. J. 2001 Drag reduction by additives: a review. In *Turbulence Structure and Modulation* (ed. A. Soldati & R. Monti), *CISM*, vol. 415, pp. 269–316. New York, NY: Springer-Verlag.

Nikuradse, J. 1932 Gesetzmässigkeiten der turbulenten Strömung in glatten Rohren. Forschungsheft 356. VDI.

Ooi, A., Martin, J., Soria, J. & Chong, M. S. 1999 A study of the evolution and characteristics of the invariants of the velocity-gradient tensor in isotropic turbulence. *J. Fluid Mech.* **381**, 141–174.

Panchapakesan, N. R. & Lumley, J. L. 1993 Turbulence measurements in axisymmetric jets of air and helium. part 1. air jet. *J. Fluid Mech.* **246**, 197–223.

Pedlosky, J. 1987 *Geophysical Fluid Dynamics*. New York, NY: Springer.

Perry, A. E., Henbest, S. & Chong, M. S. 1986 A theoretical and experimental study of wall turbulence. *J. Fluid Mech.* **165**, 163–199.

Pope, S. B. 2000 *Turbulent Flows*. Cambridge, UK: Cambridge University Press.

Proudman, J. 1916 On the motion of solids in a liquid possessing vorticity. *Proc. R. Soc. Lond. A* **92**, 408–424.

Ptasinski, P. K., Boersma, B. J., Nieuwstadt, F. T. M., Hulsen, M., Brule, B. H. A. A. & Hunt, J. C. R. 2003 Turbulent channel flow near maximum drag reduction: simulations, experiments and mechanics. *J. Fluid Mech.* **490**, 251–291.

Ptasinski, P. K., Nieuwstadt, F. T. M., Brule, B. H. A. A. & Hulsen, M. 2001 Experiments in turbulent pipe flow with polymer additives at maximum drag reduction. *Flow Turbul. Combust.* **66**, 159–182.

Rajaratnam, N. 1976 *Turbulent Jets*. Amsterdam: Elsevier.

Reeks, M. W. 1983 The transport of discrete particles in inhomogeneous turbulence. *J. Aerosol. Sci.* **14**, 729–739.

Reynolds, O. 1883 An experimental investigation of the circumstances which determine whether the motion of water shall be direct or sinuous, and of the law of resistance in parallel channels. *Phil. Trans. R. Soc.* **174**, 935.

Reynolds, O. 1895 On the dynamical theory of incompressible viscous fluids and the determination of the criterion. *Phil. Trans. R. Soc.* **186**, 123–164.

Richardson, L. F. 1922 *Weather Prediction by Numerical Process*, 2nd edn. Cambridge, UK: Cambridge University Press.

Rodi, W. 1975 A new method of analyzing hot wire signals in highly turbulent flow and its evaluation in a round jet. *DISA Information* **17**.

Rutgers, M. A. 1998 Forced 2D turbulence: experimental evidence of simultaneous inverse energy and forward enstrophy cascades. *Phys. Rev. Lett.* **81**, 2244–2247.

Saddoughi, S. G. & Veeravalli, S. V. 1994 Local isotropy in turbulent boundary layers at high reynolds number. *J. Fluid Mech.* **268**, 333–372.

Saffman, P. G. 1965 The lift on a small sphere in a slow shear flow. *J. Fluid Mech.* **22**, 385–400.

Sagaut, P. 2005 *Large Eddy Simulation for Incompressible Flow*. Berlin: Springer-Verlag.

Schlatter, P., Li, Q., Brethouwer, G., Johansson, A. V. & Henningson, D. S. 2010 Simulations of spatially evolving turbulent boundary layers up to $Re_\theta = 4300$. *Int. J. Heat Fluid Fl.* **31**, 251–261.

Schlichting, H. 1979 *Boundary-Layer Theory*, 7th edn. New York, NY: McGraw-Hill.

Schmid, P. J. 2007 Nonmodal stability theory. *Annu. Rev. Fluid Mech.* **39**, 129–162.

Schuster, H. G. 1984 *Deterministic Chaos*. Weinheim, D: Physik-Verlag.

Schuster, H. G. & Just, W. 2005 *Deterministic Chaos*, 4th edn. Wiley-VCH.

Shockling, M. A., Allen, J. J. & Smits, A. J. 2006 Roughness effects in turbulent pipe flow. *J. Fluid Mech.* **564**, 267–285.

Simonin, O., Deutsch, E. & Minier, J. P. 1993 Eulerian prediction of the fluid/particle correlated motion in turbulent two-phase flows. *Appl. Sci. Res.* **51**, 275–283.

Smith, C. R. 1984 A synthesized model of the near-wall behavior in turbulent boundary layers. In *Proc. 8th Symp. on Turbulence* (ed. G. K. Patterson & J. L. Zakin). Rolla, MO.

Spalart, P. R. 1988 Direct simulation of a turbulent boundary layer up to $R_\theta = 1410$. *J. Fluid Mech.* **187**, 61–98.

Sreenivasan, K. R. 1991 Fractals and multifractals in fluid turbulence. *Annu. Rev. Fluid Mech.* **23**, 539–600.

Staplehurst, P. J., Davidson, P. A. & Dalziel, S. B. 2008 Structure formation in homogeneous freely decaying rotating turbulence. *J. Fluid Mech.* **598**, 81–105.

Sureshkumar, R., Beris, A. N. & Handler, R. A. 1997 Direct numerical simulation of the turbulent channel flow of a polymer solution. *Phys. Fluids* **9**, 743–755.

Taylor, G. I. 1917 Motion of solids in fluids when the flow is not irrotational. *Proc. R. Soc. Lond. A* **93**, 99–113.

Taylor, G. I. 1938 The spectrum of turbulence. *Proc. Roy. Soc. London A* **132**, 476–490, also: G.I. Taylor, Scientific Papers Vol. II (Ed. G.K. Batchelor), CUP.

Tennekes, H. & Lumley, J. L. 1972 *A First Course in Turbulence*. Cambridge: MIT Press.

Toms, B. A. 1948 Some observations on the flow of linear polymer solutions through straight tubes at large Reynolds numbers. In *Proc. 1st. Int. Congr. Rheol.*, vol. 2, pp. 135–141. Amsterdam: North Holland.

Tong, C. & Warhaft, Z. 1995 Passive scalar dispersion and mixing in a turbulent jet. *J. Fluid Mech.* **292**, 1–38.

den Toonder, J. M. J. & Nieuwstadt, F. T. M. 1997 Reynolds number effects in a turbulent pipe flow for low to moderate Re. *Phys. Fluids* **9**, 3398–3409.

Townsend, A. A. 1976 *The Structure of Turbulent Shear Flow*, 2nd edn. Cambridge University Press.

Tropea, C., Yarin, A. & Foss, J., ed. 2007 *Handbook of Experimental Fluid Mechanics*. Springer.

Uberoi, M. S. & Freymuth, P. 1969 Spectra of turbulence in wakes behind circular cylinders. *Phys. Fluids* **12**, 1359–1363.

van Doorne, C. W. H. & Westerweel, J. 2007 Measurement of laminar, transitional and turbulent pipe flow using stereoscopic-PIV. *Exp. Fluids* **42**, 259–279.

van Doorne, C. W. H. & Westerweel, J. 2009 The flow structure of a puff. *Phil. Trans. R. Soc. A* **367**, 489–507.

Van Dyke, M. 1982 *An Album of Fluid Motion*. Stanford, CA: Parabolic.

van Haarlem, B., Boersma, B. J. & Nieuwstadt, F. T. M. 1998 Direct numerical simulation of particle deposition onto a free-slip and no-slip surface. *Phys. Fluids* **10**, 2608–2620.

Vincent, A. & Meneguzzi, M. 1991 The spatial structure and statistical properties of homogeneous turbulence. *J. Fluid Mech.* **225**, 1–20.

Virk, P. S. 1975 Drag reduction fundamentals. *AIChE J.* **21**, 625–656.

Virk, P. S., Merrill, E. W., Mickley, H. S., Smith, K. A. & Mollo-Christensen, E. L. 1967 The Toms phenomenom: turbulent pipe flow of dilute polymer solutions. *J. Fluid Mech.* **30**, 305–328.

Vuorinen, V., Wehrfritz, A., Yu, J., Kaario, O., Larmi, M. & Boersma, B. J. 2011 Large-eddy simulation of subsonic jets. *J. Phys. Conf. Ser.* **318**, 032052.

Waleffe, F. 1997 On a self-sustaining process in shear flows. *Phys. Fluids* **9**, 883–900.

Wallace, J. M. & Foss, J. F. 1995 The measurement of vorticity in turbulent flows. *Annu. Rev. Fluid Mech.* **27**, 469–514.

Wedin, H. & Kerswell, R. R. 2004 Exact coherent structures in pipe flow: travelling wave solutions. *J. Fluid Mech.* **508**, 333–371.

Westerweel, J., Draad, A. A., Van der Hoeven, J. G. Th. & Van Oord, J. 1996 Measurement of fully-developed turbulent pipe flow with digital particle image velocimetry. *Exp. Fluids* **20**, 165–177.

Westerweel, J., Elsinga, G. E. & Adrian, R. J. 2013 Particle image velocimetry for complex and turbulent flows. *Annu. Rev. Fluid Mech.* **45**, 369–396.

Westerweel, J., Fukushima, C., Pedersen, J. M. & Hunt, J. C. R. 2009 Momentum and scalar transport at the turbulent/non-turbulent interface of a jet. *J. Fluid Mech.* **631**, 199–230.

White, C. M. & Mungal, M. G. 2008 Mechanics and prediction of turbulent drag reduction with polymer additives. *Annu. Rev. Fluid Mech.* **40**, 235–256.

White, F. M. 1991 *Viscous Fluid Flow*, 2nd edn. McGraw-Hill.

White, F. M. 2011 *Fluid Mechanics*, 7th edn. New York, NY: McGraw-Hill.

Whitham, G. B. 1974 *Linear and Nonlinear Waves*. New York, NY: Wiley.

Wu, X. & Moin, P. 2008 A direct numerical simulation study on the mean velocity characteristics in turbulent pipe flow. *J. Fluid Mech.* **608**, 81–112.

Wu, X. & Moin, P. 2009 Direct numerical simulation of turbulence in a nominally zero-pressure-gradient flat-plate boundary layer. *J. Fluid Mech.* **630**, 5–41.

Wygnanski, I. & Fiedler, H. E. 1969 Some measurements in the self-preserving jet. *J. Fluid Mech.* **38**, 577–612.

Wygnanski, I. J. & Champagne, F. H. 1973 On transition in a pipe. part 1. the origin of puffs and slugs and the flow in a turbulent slug. *J. Fluid Mech.* **59**, 281–335.

Wyngaard, J. C., Coté, O. R. & Izumi, Y. 1971 Local free convection, similarity, and the budgets of shear stress and heat flux. *J. Atmos. Sci.* **28**, 1171–1182.

Yoshimatsu, K., Midorikawa, M. & Kaneda, Y. 2011 Columnar eddy formation in freely decaying homogeneous rotating turbulence. *J. Fluid Mech.* **677**, 154–178.

Zagarola, M. V. & Smits, A. J. 1998 Mean-flow scaling of turbulent pipe flow. *J. Fluid Mech.* **373**, 33–79.

Index

A

Added mass, *see* Particle
Advection, 9, 55, 194, 226, 247
    equation (nonlinear), 56, *57*, 59
Aeroacoustics, **244**, *see also* Sound
Alaska pipe line, 257
Algebraic stress model, *see* Closure
Aliasing, *see* Spectrum
Anisotropy, 134, 159, 167, *169*, 175, 203
    rotation, 249, 254
Atmosphere, *see* Boundary layer
Attractor, *see* Dynamical system
Austausch coefficient, *see* Eddy viscosity
Average, *see* Statistics

B

Bénard convection, *see* Convection
Bandwidth, *see* Spectrum
Basset history force, *see* Particle
Batchelor spectrum, *see* Spectrum
Bernoulli's law, 23
Bifurcation, *see* Stability analysis
Blasius friction law, *see* Pipe flow
Boundary, *see* Interface
    condition, 2, 22, 87, 114, 117, 137, 220,
        238, 243, 255
        dynamic -, 22
        ideal -, 3
        kinematic -, 22
        periodic -, 71
    layer, *see* Boundary layer
Boundary layer, 33, *34*, *81*, 95, *95*, 100, 104,
        106, *157*, *160*, 200, 235, 240, *240*,
        252
    atmospheric -, 15, 16, 235
        inversion, 146

    neutral -, 16, 236
        stable -, 16, *16*, 236
        unstable -, 16, 236
    convective -, *see* Convection
    closure, *see* Closure
    defect law, *97*, *see also* Wall turbulence
    Ekman -, *see* Rotating turbulence
    equations, 95, 107, 109, 113
    pressure gradient, *see* Pressure
    separation, 105, *106*, *see also* Pressure
        gradient
    shape factor, 106
    thickness, 104, 146
        displacement -, 105
        momentum loss -, 106
    transition, 34, 42
    visualization of -, *96*
Boussinesq
    approximation, **12**, 13, 24, 26, 80, 143
    closure, *see* Closure
    equations, 14, 19, 138, **235**
        rotating frame of reference, 257
    reference state, 13
Brunt–Väisälä frequency, 16, 31
Buffer layer, *see* Wall turbulence
Buoyancy, 139, 224, 236
    Morton length, 123
    production, *see* Kinetic energy
Burgers equation, 6, **55**, 62, 130, 221
    dimensionless -, 58
    solution, 58, 59, *59*
Burst, *see* Intermittency, *see* Structure (turbulent)
Bypass transition, *see* Transition

© Springer International Publishing Switzerland 2016
F.T.M. Nieuwstadt et al., *Turbulence*, DOI 10.1007/978-3-319-31599-7

Printed in the United States
By Bookmasters